DICKIE, DONALD E
LIFTING TACKLE MANUAL
000457493

621.861 D55

LIFTING TACKLE MANUAL

LIFTING TACKLE MANUAL

Compiled by
D. E. DICKIE, P.Eng.
*Senior Engineer, Equipment Research
Research and Development Department
Construction Safety Association of Ontario*

United Kingdom Editor
DOUGLAS SHORT, B.Sc., M.I.C.E., Hon.F.I.I.S.O.

BUTTERWORTHS
London Boston Sydney Wellington Durban Toronto

THE BUTTERWORTH GROUP

United Kingdom Butterworth & Co (Publishers) Ltd
London: 88 Kingsway, WC2B 6AB

Australia Butterworths Pty Ltd
Sydney: 586 Pacific Highway, Chatswood, NSW 2067
Also at Melbourne, Brisbane, Adelaide and Perth

New Zealand Butterworths of New Zealand Ltd
Wellington: 31–35 Cumberland Place,
CPO Box 472

South Africa Butterworth & Co (South Africa) (Pty) Ltd
Durban: 152–154 Gale Street

First published in Canada 1975
by Construction Safety Association of Ontario
Revised UK edition published 1981
by Butterworths

© Construction Safety Association of Ontario, 1975

British Library Cataloguing in Publication Data
Dickie, D E
   Lifting Tackle Manual. – Revised and metricated ed.
   1. Hoisting machinery – Safety measures
   I. Title   II. Short, Douglas
   III. Construction Safety Association of
   Ontario   IV. Rigging Manual
   621.8'7   TJ1350

ISBN 0-408-00446-0

All rights reserved. No part of this publication may be reproduced or transmitted in any form or by any means, including photocopying and recording, without the written permission of the copyright holder, application for which should be addressed to the Publishers. Such written permission must also be obtained before any part of this publication is stored in a retrieval system of any nature.

This book is sold subject to the Standard Conditions of Sale of Net Books and may not be resold in the UK below the net price given by the Publishers in their current price list.

Typeset by Butterworths Litho Preparation Department
Printed in England by Butler & Tanner Ltd., Frome and London

Preface

This manual, and its companion volume *Crane Handbook*, were first published in Ontario, Canada, in October 1975 and quickly attracted sufficient attention to justify special editions for the United Kingdom and the Commonwealth countries.

The books give practical advice without making detailed reference to government or state safety regulations but, in order to avoid possible confusion or misunderstanding, the UK editions contain lists of principal statutory requirements in the UK, the relevant British Standards, and the addresses of the area offices of the Health & Safety Executive which administer the statutory requirements.

Editions in French were published by the Government of Quebec in March 1978.

This manual applies to lifting tackle, materials, procedures and precautions used in the construction industry.

The Construction Safety Association of Ontario believes that proper job training is the answer to improved accident prevention, and this manual is an attempt to identify and expand upon the principles of use of lifting tackle by prescribing minimum safety requirements, guidelines and procedures. It should be used as a guide in conjunction with the applicable safety regulations by contractors, supervisors, slingers, crane drivers and employees who are concerned with or responsible for construction safety.

The text has been endorsed by both labour and management authorities in Ontario and is fully a document of accord.

In offering this manual to the industry, the Construction Safety Association of Ontario hopes to contribute to a broader understanding of lifting tackle and its importance in the construction field. It is intended that it should become a working guide and that the information be used in the training of personnel involved with lifting tackle and be included in standing instructions for the safe use of this equipment.

Knowledge of the equipment and materials with which we work is one of the most important factors in accident prevention. Each piece of equipment and material has been designed and developed to serve a specific purpose and knowledge of what it can and cannot do not only improves efficiency but also eliminates hazards.

Owing to the many variations in lifting tackle practices and the different ways in which this tackle is used, recommendations must, of necessity, be framed in general terms. The recommendations can only be advisory in nature and are intended to complement relevant regulations and manufacturers' requirements which must be observed.

In the UK cranes and lifting tackle are subject to the general safety requirements of the Health and Safety at Work Act, 1974 and the more detailed ones of the Construction (Lifting Operations) Regulations, 1961. One requirement is that there must be an official certificate of test for every crane or item of lifting tackle.

Both metric and Imperial units are shown in this manual except where there is a clear relation between the figures as in the 'tonne' and the 'ton'. British Standards' recommendations have been adopted in place of Canadian ones but where the latter provide practical advice which is not readily available from the former, the information in the Canadian edition has been retained.

The Construction Safety Association of Ontario is grateful to and wishes to thank all who assisted in the preparation of this manual. Special thanks go to the following for their contributions and capable assistance:

> British Ropes Ltd.
> Greening Donald Co. Ltd.
> Johnson Blocks, Div. of D.R. Hinderliter Inc.
> Martin Black Wire Ropes of Canada Ltd.
> Mr. R. B. Martin, P.Eng., Ontario Hydro
> Ontario Hydro
> The Crosby Group
> The Workers' Compensation Board of British Columbia
> Wire Rope Industries of Canada Ltd.

L. Sylvester,
General Manager.

G. J. Samson,
Executive Director.

CONSTRUCTION SAFETY ASSOCIATION OF ONTARIO

(Douglas Short, editor of the United Kingdom edition, died while engaged on checking the final draft before publication)

Note: The contents of this manual and of its companion volume, *Crane Handbook*, are subject to the laws and regulations in force in any territory in which they may be used.

CONTENTS

1. WIRE ROPE — 1
SELECTION — 1
Design and construction — 2
Grades — 3
Strand classifications — 3
Constructions — 3
Rope lay — 5
Preforming — 5
Cores — 8
Size — 9
Fatigue and abrasion resistance — 10
Safety factors — 11
Safe working loads (SWL) — 12
Non-rotating ropes — 13
INSTALLATION — 17
INSPECTION — 25
USE, HANDLING AND MAINTENANCE — 39
Procedures and precautions — 39
Lubrication — 40
Storage — 43
Seizing and cutting — 43
Splicing — 45
END FITTINGS AND CONNECTIONS — 45

2. FIBRE ROPE — 55
TYPES AND CHARACTERISTICS — 55
Natural fibre ropes — 55
Synthetic fibre ropes — 57
Comparison of ropes — 60
Safety factors — 64
Safe working loads (SWL) — 65
CARE AND USE OF FIBRE ROPE — 67
Coiling and uncoiling — 67
Whipping — 67
Storage — 68
Use — 69
INSPECTION — 70
Splicing — 74
Splices — 74
KNOTS, BENDS AND HITCHES — 77

3. CHAIN — 89
Grades — 89
Strength — 90
Inspection and examination — 90
Care and use — 93

4. OTHER EQUIPMENT — 94
Drums — 94
Sheaves — 97
Hooks — 103
Rings, links, swivels — 106
Shackles — 106
Eye bolts — 109
Turnbuckles — 110
Spreader and equaliser beams — 111
Blocks — 112

5. REEVING — 120

6. SLINGS — 131
Sling configurations — 131
Sling angles — 134
Safe working loads (SWL) — 136
Fibre rope slings — 140
Synthetic webbing slings — 146
Metal (wire or chain) mesh slings — 150
Chain slings — 150
Wire rope slings — 155

7. OPERATING PROCEDURES AND PRECAUTIONS — 169
Responsibilities — 169
Procedures and precautions — 169
Determining load weights — 180
Centre of gravity (C of G) — 181
Signalling — 183

APPENDICES
  A British Standards and principal statutory requirements — 187
  B HSE area offices and other addresses — 188

TABLES

1. WIRE ROPE — 1
- 1.1 Service requirements and design characteristics — 2
- 1.2 Diameter tolerances — 9
- 1.3–8 Safe working loads (SWL) — 12–15
- 1.9 Faults/possible causes — 37
- 1.10 Recommended number of seizings and size of seizing wire — 44
- 1.11 Rope diameter/efficiency — 49
- 1.12 Installation of wire rope clips — 51
- 1.13 Installation of double saddle clips — 54

2. FIBRE ROPE — 55
- 2.1 Relative strength of fibre ropes (dry) — 60
- 2.2 % change in rope strength when wet — 60
- 2.3 Energy absorption capacity — 61
- 2.4 Influence of sustained loading — 61
- 2.5 Rope performance under repeated loading and unloading — 62
- 2.6 Relative flexing and bending endurance — 63
- 2.7 Effect of hot surfaces — 63
- 2.8 Effect of radiated heat and hot gases — 63
- 2.9 Strength loss v temperature — 63
- 2.10 Effect of low temperature on rope strength — 64
- 2.11 Approximate SWL of new fibre ropes — 66
- 2.12 Approximate SWL of new braided synthetic fibre ropes — 66

3. CHAIN — 89
- 3.1 SWL of alloy steel chain — 89
- 3.2 Effect of heat on SWL — 90
- 3.3 Correction table to compensate for wear — 90

4. OTHER EQUIPMENT — 94
- 4.1 Drum/reel capacity factors — 97
- 4.2 Contact pressure limits for sheave and drum materials — 99
- 4.3 Drum diameters for various types of rope — 99
- 4.4 Sheave groove tolerances — 101
- 4.5 SWL of eye hooks, shank hooks, swivel hooks — 105
- 4.6 SWL of chain slip hooks (clevis type and eye type) — 105
- 4.7 SWL of rings — 105
- 4.8 SWL of links alternative to rings — 105
- 4.9 SWL of reevable egg links — 105
- 4.10 SWL of intermediate links — 106
- 4.11 SWL of swivels — 106
- 4.12 SWL of shackles (dee/bow) — 107
- 4.13 SWL of eye bolts — 107
- 4.14 SWL of turnbuckles — 112
- 4.15 Multiplication factors for snatch block loads — 116

5. REEVING — 120
- 5.1–3 Factors to acount for sheave friction loads — 120–123

6. SLINGS — 131
- 6.1 Effect of sling angle measurement error on loads — 136
- 6.2 SWL of manila rope slings — 142

6.3 SWL of nylon rope slings	143
6.4 SWL of polypropylene rope slings	144
6.5 SWL of polyester rope slings	145
6.6 SWL of nylon web slings 1400 kN/m (8000 lb/in) material	151
6.7 SWL of nylon web slings 1050 kN/m (6000 lb/in) material	152
6.8 SWL of Dacron web slings 875 kN/m (5000 lb/in) material	153
6.9 SWL of metal (wire or chain) mesh slings	156
6.10 SWL of chain slings (alloy steel)	157
6.11–14 SWL of wire rope slings	158–161
6.15 SWL of braided wire rope slings (6-part braided rope)	162
6.16 SWL of braided wire rope slings (8-part braided rope)	163
6.17 SWL of cable-laid wire rope slings	164
6.18 SWL of strand-laid grommet slings	165
6.19 SWL of cable-laid grommet slings	166
6.20 SWL of strand-laid endless slings	167
6.21 SWL of cable-laid endless slings	168

7. OPERATING PROCEDURES AND PRECAUTIONS — 169

7.1 Approximate weights of round steel bars and rods	181
7.2 Weights of materials (based on volume)	183

CHAPTER 1

Wire rope

SELECTION

The responsibility of equipment selection and application involves using material that will not only get the job done as quickly and economically as possible, but that eliminates all possibility of hazard to personnel on the site, the public and the property for as long as it will be used and under all anticipated conditions to which it will be exposed and operated.

Nothing can take the place of experience in making these decisions, however, but the guidelines set out in this manual are intended to simplify the process by stressing those critical considerations that must not be overlooked.

It is both the equipment supplier's and the contractor's responsibility to ensure that these requirements are met.

Many factors influence the selection of wire rope. The strength of the rope, although of major importance, is only one of these factors. It is advisable, therefore, to use only ropes of the correct size, grade, type and construction as specified by either the equipment manufacturer or the rope manufacturer who must base their recommendations on actual working conditions.

There are six requirements which must be considered when selecting a wire rope:

(1) The rope must possess sufficient strength to take the maximum load that may be applied, with in general a factor of safety (FoS) of at least 6:1 for materials and 12:1 when carrying personnel.

The wire ropes that are fitted to mobile cranes must possess factors of safety as follows:

– live or running ropes that wind on drums or pass over sheaves.
   = 4.5–5.5:1 depending on crane class and duty
   = 3.75:1 when erecting the jib
– pendants or standing ropes
   = 3.5–4:1 under operating conditions
   = 3:1 when erecting the jib.

Note: With regard to these figures, see BS.302 *Wire Ropes for Cranes etc.*, BS.1757 *Power-driven Mobile Cranes*, and BS.2799 *Power-driven Tower Cranes*.

(2) The rope must withstand repeated bending without failure of the wires from fatigue.
(3) The rope must resist abrasion.
(4) The rope must withstand distortion and crushing.
(5) The rope must resist rotation.
(6) The rope must resist corrosion.

Table 1.1 illustrates how the design and construction of the rope can be varied to give the best characteristics for specific service demands.

TABLE 1.1

WIRE ROPE SERVICE REQUIREMENTS	WIRE ROPE DESIGN CHARACTERISTICS
Strength: the wire rope must develop sufficient strength to support the load plus the necessary factor of safety.	The strength of a wire rope depends on its size, grade of wire and type of core.
Flexibility, or resistance to bending fatigue: the wire rope must have the ability to bend over small sheaves or wind onto relatively small drums without the wires breaking due to bending fatigue.	Strands containing a large number of small wires have greater resistance to bending fatigue than strands containing a few large wires. Langs lay has greater fatigue resistance than regular lay. Preforming increases the wire rope's resistance to bending fatigue.
Resistance to abrasion: the wire rope is subjected to wear or abrasion as it passes through operating sheaves under high pressure or comes in contact with stationary objects.	Large outer wires are better able to withstand abrasive wear. Langs lay provides greater resistance to wear than regular lay.
Resistance to crushing: some wire ropes distort or flatten when they are forced to operate under heavy pressure in grooves that do not provide ample support or on drums where multiple lay winding occurs.	An independent wire rope core (IWRC) provides greater support for strands under heavy bearing pressures. The coarser wire rope constructions provide greater resistance to flattening on drums.
Resistance to rotation: a wire rope may rotate as the load is applied. This could be undesirable for load control and might lead to rapid deterioration of the wire rope.	Special non-rotating constructions are available for specific applications. Regular lay provides greater stability than Langs lay, and wire ropes with an IWRC twist less than those with fibre cores.
Resistance to corrosion: wire rope may corrode if in contact with corrosive elements or it may rust when exposed to atmospheric conditions over a long period of time.	Galvanised or stainless steel wires offer excellent protection against corrosion. Special lubricants can also inhibit the development of rust.

DESIGN AND CONSTRUCTION

Wire rope consists of many individual wires laid into a number of strands which are, in turn, laid around a centre core. (Fig. 1.1).

The type and size of wire used, the number of wires in the strands, and the type of core determine the strength of a wire rope of a given size. A typical description of a wire rope would be: 300 m (1000 ft), 13 mm (½ in) diameter, 6 × 19 (9/9/1), preformed, 180 (115 tonf/in^2), fibre core, Langs lay.

The precise meaning of each term is as follows:

(a) 300 m (1000 ft) – the length of wire rope is ordered or recorded in metres or feet.
(b) 13 mm (½ in) diameter – the nominal diameter of the rope.
(c) 6 × 19 (9/9/1) – the first numeral is the number of strands in the rope (6). The second numeral is the number of wires in each strand (19). The set of numerals in ( ) shows the pattern of wires in the strand.
(d) Preformed – a type of process assuring that each strand of the rope is preformed to the helical shape it will assume in the finished rope.
(e) The figure 180 shows the tensile grade of the wire.

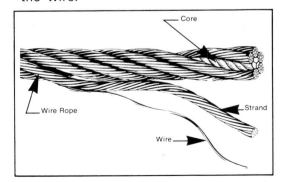

Fig. 1.1 Composition of wire rope

WIRE ROPES

(f) 'Fibre core' indicates the type of core in the rope.
Note: With regard to (e) and (f), see BS.302 *Wire Ropes for Cranes etc.*, and BS.2763 *Round Steel Wire for Wire Ropes*.
(g) Langs lay — indicates the direction of the strands and the wires in the strands.

Each distinct design element influences the rope's characteristics. The design objective is to provide a rope with characteristics balanced to withstand the demands of a particular service. There are five basic design elements:

(1) Grade of wire.
(2) Number and pattern of wires in the strand.
(3) Type of lay.
(4) Preforming.
(5) Type of core.

Each of these elements will be explained in more detail since a thorough understanding of wire ropes contributes to lifting and lifting tackle safety.

GRADES

To meet the demand for varying degrees of strength, toughness, abrasion resistance, and corrosion resistance, wire rope is manufactured in the following grades for general crane and lifting operations:

(1) *Grade 200 (130 tonf/in^2):* Used for special installations where maximum rope strength is required.
(2) *Grade 180 (115 tonf/in^2):* Used for special applications where breaking strengths somewhat higher than obtained with Grade 160 are desired and where other conditions such as sheave and drum diameters are favourable to its use.
(3) *Grade 160 (102 tonf/in^2):* Has a good combination of tensile strength, wearing qualities and excellent fatigue-resistance properties. These steels have high strength, are tough and ductile, and have high wear resistance.
(4) *Grade 145 (92 tonf/in^2):* Has a lower tensile strength and resistance to wear than Grade 160 but retains high fatigue-resistance properties. Can be used when stength is secondary to wear resistance.

STRAND CLASSIFICATIONS

Wire ropes are also characterised by their strand shape (Fig. 1.2) as follows:

(1) Round strand ropes.
(2) Flattened strand ropes.
(3) Locked coil ropes.
(4) Concentric strand ropes.

Since almost all lifting and lifting tackle ropes are of round strand classification the discussion in this manual will be confined to them.

CONSTRUCTIONS

Round strand ropes are grouped into classifications according to the number of wires per strand. As the number increases, the rope develops greater ability to resist bending fatigue — at the expense of resistance to abrasion.

Within the round strand classification, the wires may be arranged in certain geometric patterns, called 'constructions'. The number and pattern of the wires define the rope's characteristics.

The wires in the strands may be all the same size or a mixture of sizes (Fig. 1.3). There are four common arrangements:

(1) *Ordinary:* where the wires are all the same size.
(2) *Seale:* where wires of larger diameter are used on the outside of the strand to resist abrasion and smaller wires are inside to provide flexibility.
(3) *Warrington:* where alternate wires are large and small to combine great flexibility with resistance to abrasion.
(4) *Filler:* where very small wires fill in the valleys between the outer and inner rows of wires to provide good resistance to abrasion and fatigue.

As the number of wires per strand increases, combinations of filler, Seale, and Warrington patterns are used to provide certain characteristics. This combination of rope classifications and strand constructions produces varying degrees of resistance to bending fatigue and abrasion. The effect of varying constructions can be best shown by using, as examples, the most commonly-used ropes, the 19-wire strand (Fig. 1.4), and 37-wire strand ropes (Fig. 1.5), as follows:

Basic 19-wire strands:

— *19-wire Seale:* Seale patterns always have the same number of outer and inner wires.

WIRE ROPES

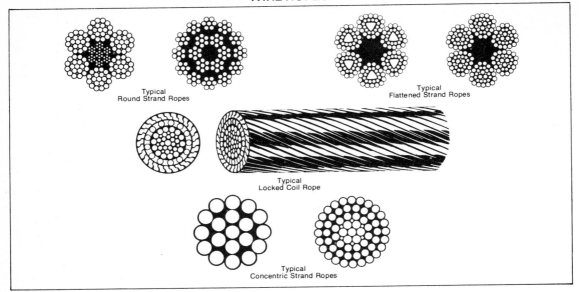

Fig. 1.2 Strand classifications

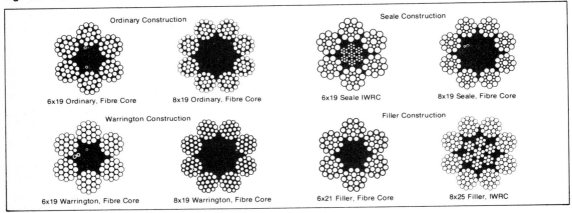

Fig. 1.3 Wire rope constructions

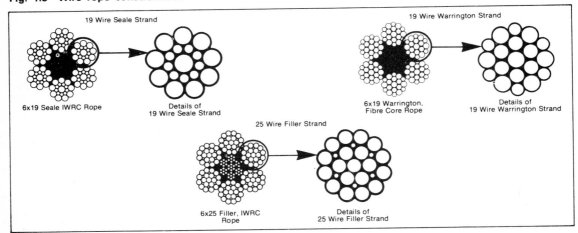

Fig. 1.4 Basic 19-wire-strand construction

The outer wires are fairly large and provide good abrasion resistance.
- *19-wire Warrington:* Outer wires are two sizes and are alternately supported by the crowns and valleys of the inner layer. This type of construction is ideal for small-diameter ropes or for intermediate layers when building large diameter strands of many wires.
- *25-wire filler:* Filler wires are used to prevent the outer wires from crowding into the valleys between the inner wires. The filler construction has good abrasion and fatigue-resistance characteristics.

Basic 37-wire strands:

- *31-wire Warrington Seale:* In this built-up construction, the 19-wire Warrington strand is enclosed by an outler layer of 12 large-diameter wires resting in the valleys between the 12 wires of the inner layer. This construction provides abrasion resistance similar to 25-wire filler, but increased flexibility from additional internal wires.
- *41-wire filler:* The 41-wire filler comprises a 17-wire inner Seale strand, with an outer layer of 16 wires and a layer of eight filler wires to prevent crowding of the outer wires. This construction is recommended for larger-diameter ropes where more flexibility is required.
- *49-wire filler Seale:* The 49-wire filler Seale comprises a conventional 33-wire filler strand with an outer layer of 16 large-diameter wires resting between the 16 wires of the inner layer. This construction is ideal for large-diameter ropes where the requirement is maximum flexibility.

ROPE LAY

The term 'rope lay' signifies the direction of rotation of the wires and the strands in the rope. Rotation is either to the right (clockwise) or to the left (counterclockwise).

The lay of the rope affects its flexibility and resistance to wear. The term 'lay length' is the distance measured along a rope in which a strand makes one complete revolution around the rope axis. (Fig. 1.6)

In *regular lay ropes* (Fig. 1.7) the wires in the strands are laid in one direction while the strands in the rope are laid in the opposite direction. The result is that the wire crown runs approximately parallel to the longitudinal axis of the rope. These ropes are stable, have good resistance to kinking and twisting and are easy to handle. Due to the short length of exposed wires, they are also able to withstand considerable crushing and distortion.

In *Langs lay ropes* (Fig. 1.8) the wires in the strands and the strands in the ropes are laid in the same direction. The outer wires run diagonally across the rope and are exposed for longer lengths than in the regular lay rope. With the outer wires presenting greater wearing surfaces, Langs lay ropes have greater resistance to abrasion. They are also more flexible and possess greater resistance to fatigue. They are more liable to kinking and untwisting and are not capable of withstanding the same abuse from distortion and crushing. Langs lay ropes should have *both* ends permanently fastened to prevent untwisting and as such they should not be recommended for use on single-part hoist ropes nor should they be used with swivel end terminals.

Right lay and left lay (Fig. 1.9). A right lay rope is one in which the strands rotate or twist to the right like a conventional screw thread. A left lay rope is just the opposite. No rope should be ordered left lay without first considering all features of its proposed end use.

Alternate lay ropes (Fig. 1.10) have three strands made with right lay and three with left lay. The six strands are then positioned in the finished rope so that the strands alternate.

Herringbone or twin strand ropes (Fig. 1.11) have a combination of right-hand laid and left-hand laid strands. They usually consist of two pairs of the right-hand laid strands (Langs lay) and two single left-laid strands (regular lay). The two single left-laid strands are separated by the pairs of right-hand strands. This arrangement gives some of the stability of a regular lay rope while retaining a good portion of the bending characteristics of Langs lay.

The standard rope, unless otherwise stated, is understood to be right regular lay.

PREFORMING

The wires and strands in preformed rope are shaped in the manufacturing process to fit their position in the finished rope. This removes the tendency of the wires and strands to straighten. (Fig. 1.12)

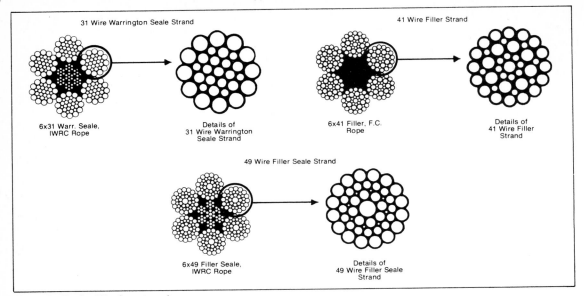

Fig. 1.5 Basic 37-wire strands

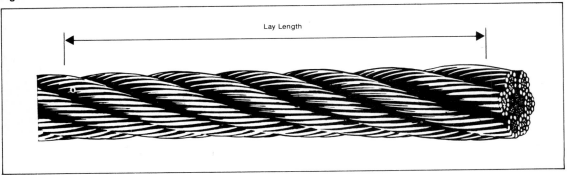

Fig. 1.6 Measurement of a rope's lay length

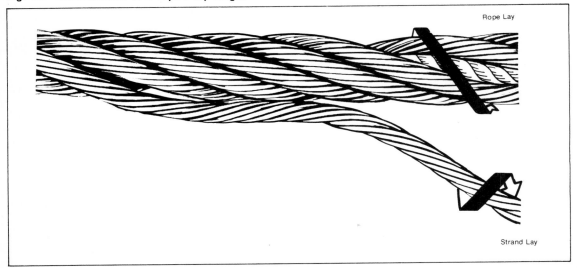

Fig. 1.7 Regular lay rope – wires and strands laid in opposite directions

WIRE ROPES

Fig. 1.8 Langs lay rope – wires and strands laid in the same direction

Fig. 1.10 Alternate lay

Fig. 1.11 Herringbone or twin-strand lay

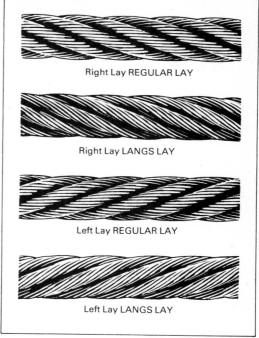

Fig. 1.9 Rope lay

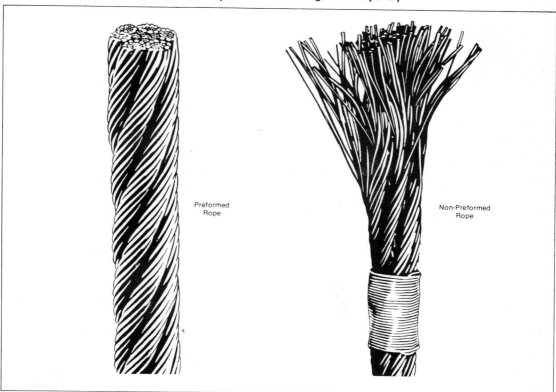

Fig. 1.12 How preforming affects wire rope

As a result:

- They can be cut without seizings.
- Broken rope ends do not untwist.
- They are free from liveliness and twisting tendencies, thus making installation and handling easier.
- They are less likely to kink or foul.
- They have increased resistance to bending fatigue.
- Each strand carries an equal share of the load.
- They cause less wear on sheaves and drums.
- Individual broken wires do not jut out from the rope when broken. They lie flat with the broken ends only slightly separated. For this reason, preformed rope requires very careful inspection.

CORES

The core forms the heart of the rope and is the component around which the main rope strands are laid. The core supports the strands and is intended to keep them from jamming against or contacting each other under normal loads and flexings.

The core may take one of several forms, depending on the conditions under which the rope will be used.

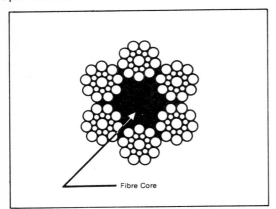

Fig. 1.13 Fibre-cored ropes

Fibre cores (Fig. 1.13) are adequate for many types of service by providing maximum flexibility and elasticity to the wire rope. While generally made of hard fibres, usually sisal and occasionally manila, they may also be manufactured from man-made fibres such as polypropylene or nylon. These synthetic cores are less susceptible to compacting, especially under moist conditions and are impervious to many acids. Do not use a fibre-core rope where it could be exposed to temperatures at which fibre cores would be damaged. Rope having an independent wire rope or wire strand core, or other temperature damage resistant core, must be used. Also, fibre core ropes should not be used when multi-layer winding is used since they are susceptible to crushing.

Wire cores (independent wire rope core, steel strand core, armoured core) (Fig. 1.14) increase the resistance of wire rope to abuse as

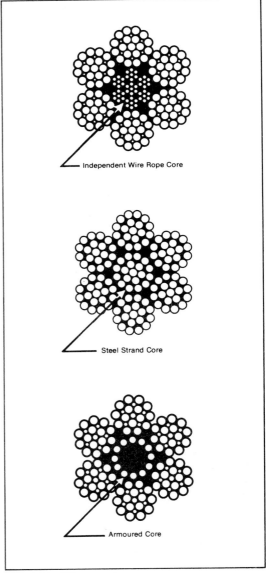

Fig. 1.14 Wire-cored ropes

WIRE ROPES

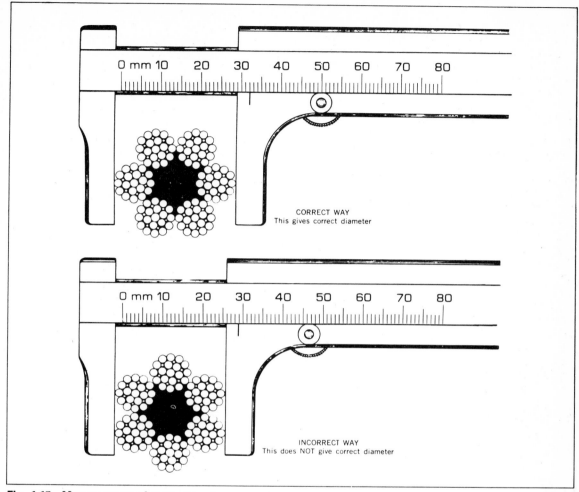

Fig. 1.15 Measurement of rope diameter

they do not yield to the compressive action of the outer strands as do fibre cores. This tends to preserve the circular cross-section of the rope when it is crushed by over-winding or when it is bent around small sheaves and drums while heavily loaded. They also prevent the strands from bridging (bearing forcibly against each other) which can result in fatigue failure. Wire cores stretch less and add to the strength of ropes (approximately 7½%) but they are less resilient than fibre-core types and are less resistant to shock loads.

SIZE

The size of a wire rope is identified by its length in metres and its diameter in millimetres. The diameter is the distance measured between the widest points (Fig. 1.15). It is advisable to make three separate measurements on a six-strand rope, and four measurements on an eight-strand rope.

Most new wire ropes measure slightly over their nominal diameter, as follows:

TABLE 1.2

DIAMETER TOLERANCES			
Nominal rope diameter		Oversize	
mm	in	mm	in
0–19	0–¾	+0.8	+1/32
20–28	13/16–1⅛	+1.2	+3/64
30–38	1 3/16–1½	+1.6	+1/16
39–56	1 9/16–2¼	+2.4	+3/32
58 and up	2 5/16 and up	+3.2	+⅛

WIRE ROPES

FATIGUE AND ABRASION RESISTANCE

The two most important factors to consider when selecting a wire rope are its fatigue and abrasion resistance because every rope in service is being slowly destroyed by fatigue and abrasion.

Fatigue is caused by constant bending of the rope over sheaves and around drums. When the load is applied to the rope, each wire is stretched with the result that they are drawn tightly together. As the wire rope is bent around a drum or sheave, there is a complex movement of the tightly compressed wires. This movement induces high stresses which may be continually reversing as the rope moves through a system of sheaves, resulting in failure of individual wires due to fatigue.

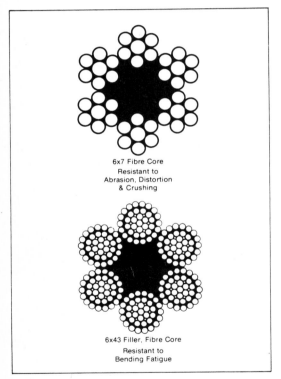

Fig. 1.16 Effect of wire size on ability to resist abrasion, distortion, crushing and bending fatigue

This point can be illustrated by bending a piece of wire back and forth at one spot; the wire will soon break due to fatigue.

This same repeated bending and straightening of the individual wires as they pass over sheaves and drums initiates small cracks in the wires that increase in number and grow larger with continued bending. The process occurs in all ropes but is accelerated if the drums and sheaves are small or if the rope is relatively inflexible.

In general, the smaller the wires and the more there are, the more flexible the rope. Adequately-sized drums and sheaves will reduce the severity of the bending stresses and increase the number of bending cycles the rope can tolerate before it is severely fatigue-weakened.

Abrasion occurs primarily on the outer surfaces of wire ropes and is caused by friction on sheaves, etc., and by dust and dirt acting as an abrasive between the rope and the sheaves and drums. Excess abrasion also occurs when the rope comes in contact with rough surfaces or sharp edges. The most difficult form of abrasion to overcome is that caused by the rotation of the wire rope as it runs over sheaves, and on drums with more than one layer of rope.

Abrasion resistance is gained by using rope having:

— Long exposed lengths of outer wires.
— Large-diameter outer wires.
— High carbon and manganese content.
— High-grade material and heat treatment in the steel.

Where abrasive conditions are severe Langs lay or specially constructed regular lay rope should be used. The wires in the Langs lay ropes are laid in the same direction as the strands: therefore greater lengths of wires on the surface or crown of the rope are in contact with the sheaves or drum than on regular lay ropes thus taking longer to wear through.

Seale rope construction has greater resistance to abrasion because it has larger outer wires which take longer to wear through. It should be noted, however, that the large wires tend to crack and break because of their great stiffness. Because of this, Seale rope should not be used unless it is specifically recommended for the job by a rope manufacturer who knows the conditions.

The ability of the rope to withstand distortion and crushing is governed principally by the use of large outer wires and wire cores to increase the resistance to abuse. (Fig. 1.16)

WIRE ROPES

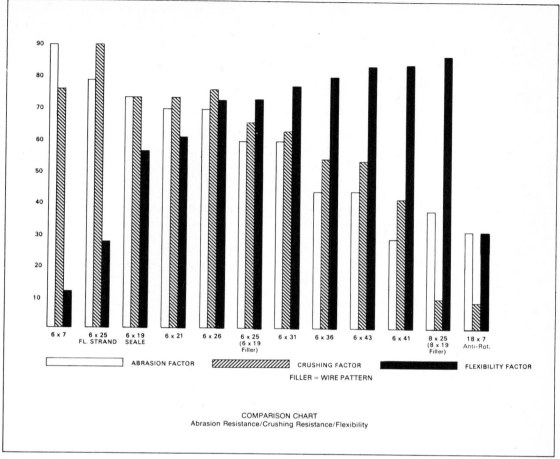

Fig. 1.17 Comparison of mechanical properties of various ropes

The ideal rope for lifting purposes provides a medium combination of these properties. For special conditions, a construction which favours the quality most necessary to the installation must be selected at a sacrifice of other qualities. (Fig. 1.17)

SAFETY FACTORS

To guard against failure of a wire rope in service, the actual load on the rope should be only a fraction of the breaking load. To account for all the stresses placed on a rope during a lifting operation and to provide the margin of strength necessary to handle loads safely and guard against accidents, it is necessary for the rope to have a Factor of Safety (FoS).

In its simplest form the FoS is defined as follows:

$$\text{Factor of Safety} = \frac{\text{Minimum Breaking Strength of the Rope}}{\text{Maximum Safe Working Load (SWL)}}$$

For lifting tackle ropes (unlike crane ropes mentioned on page 1 (1)), **the minimum acceptable factor of safety is 5, and when used on equipment that is intended to carry personnel it is 10**. See also FoS notes in relevant British Standards.

The maximum safe working load (SWL) is calculated as follows:

$$\text{Max. SWL} = \frac{\text{Minimum Breaking Strength of the Rope}}{\text{Factor of Safety}}$$

$$= \frac{\text{Minimum Breaking Strength}}{5}$$

EXAMPLE: If the wire rope catalogue gives the breaking strength of the rope as 10 tonnes, the maximum SWL is:

$$\text{Max. SWL} = \frac{10 \text{ tonnes}}{5} = 2 \text{ tonnes}$$

Too often the FoS is treated as 'reserve strength' and used for additional capacity. *It is not!*

The FoS accounts for:

- Reduced capacity of the rope below its stated breaking strength due to wear, fatigue, corrosion, abuse, and variations in size and quality.
- End fittings and splices which are not as strong as the rope itself.
- Extra loads imposed by acceleration and inertia (starting, stopping, slewing and jerking of the load).
- Increases in the load on the rope due to friction of the rope passing over sheaves.
- Inaccuracies in the weight of the load.
- Inaccuracies in the weight of the lifting tackle.
- Reduced strength of the rope due to bending over sheaves.

This list of variables is not complete. It is intended to show why a FoS is required and why it must *never* be lowered.

SAFE WORKING LOADS (SWL)

The Tables 1.3–1.8 of minimum breaking loads are based on BS.302, *Wire Ropes for Cranes and General Engineering Purposes*, and represent commonly-used types of crane rope. They relate to the most commonly-used wire rope classifications. The Tables are for reference purposes only. Check your rope manufacturer's ratings before determining the SWL as they may differ from the Tables. If the manufacturer quotes his rope's capacity in terms of breaking strength, you must divide that figure by the factor of safety to get the SWL.

$$\text{Max. SWL} = \frac{\text{Minimum breaking Strength}}{\text{Factor of Safety}}$$

TABLE 1.3

ROUND STRAND EQUAL LAY 6 × 19 (9/9/1) FIBRE CORE – BS.302				
1	2	3	4	5
Nominal diameter	Approx. equivalent diameter	Minimum breaking load at 180 kgf/mm² (115 tonf/in²)		
mm	in	kgf	kN	tonf
8	5/16	3 810	37.4	3.75
9	3/8	4 830	47.4	4.75
10	–	5 960	58.5	5.87
11	7/16	7 210	70.7	7.10
12	–	8 580	84.2	8.44
13	1/2	10 100	99.1	9.94
14	9/16	11 700	115	11.5
16	5/8	15 300	150	15.0
18	11/16	19 300	189	19.0
19	3/4	21 500	211	21.2
20	13/16	23 800	233	23.4
22	7/8	28 800	282	28.3
24	15/16	34 300	336	33.8
26	1	40 300	395	39.7
28	1 1/8	46 700	458	46.0
32	1 1/4	61 000	598	60.0
35	1 3/8	73 000	716	71.8
36	–	77 200	757	76.0
38	1 1/2	86 000	844	84.6

To obtain calculated aggregate breaking loads multiply the figures given in columns 3, 4 and 5 by 1.163.
1 kgf/mm² = 100 kgf/cm².

WIRE ROPES

Because of the difficulty in remembering the SWL of the most common wire ropes, the following rule-of-thumb should be used to obtain an approximate SWL (in tons using Imperial units). 1 ton = 1.016 tonne = 1000 kg.

SWL = ROPE DIA. × ROPE DIA. × 8

Examples:

(a) ½ in diameter rope
   SWL = ½ × ½ × 8 = 2 tons

(b) ⅝ in diameter rope
   SWL = ⅝ × ⅝ × 8 = 3.125 tons

(c) 1 in diameter rope
   SWL = 1 × 1 × 8 = 8 tons

TABLE 1.4

ROUND STRAND EQUAL LAY 6 × 19 (12/6 + 6F/1) INDEPENDENT WIRE ROPE CORE – BS.302				
1	2	3	4	5
Nominal diameter	Approx. equivalent diameter	Minimum breaking load at 180 kgf/mm² (115 tonf/in²)		
mm	in	kgf	kN	tonf
11	7/16	7 950	78.0	7.82
12	–	9 450	92.7	9.30
13	½	11 100	109	10.9
14	9/16	12 900	127	12.7
16	⅝	16 800	165	16.5
18	11/16	21 300	209	21.0
19	¾	23 700	232	23.3
20	13/16	26 200	257	25.8
22	⅞	31 800	312	31.3
24	15/16	37 800	371	37.2
26	1	44 400	436	43.7
28	1⅛	51 500	505	50.7
32	1¼	67 200	659	66.1

To obtain calculated aggregate breaking loads multiply the figures given in columns 3, 4 and 5 by 1.270.
1 kgf/mm² = 100 kgf/cm².

TABLE 1.5

TRIANGULAR STRAND 6 × 25 (12/12/△) INDEPENDENT WIRE ROPE CORE – BS.302				
1	2	3	4	5
Nominal diameter	Approx. equivalent diameter	Minimum breaking load at 180 kgf/mm² (115 tonf/in²)		
mm	in	kgf	kN	tonf
13	½	11 200	110	11.0
14	9/16	12 900	127	12.7
16	⅝	17 000	167	16.7
18	11/16	21 400	210	21.1
19	¾	24 000	235	23.6
20	13/16	26 500	260	26.1
22	⅞	32 000	314	31.5
24	15/16	38 200	375	37.6
26	1	44 700	438	44.0
28	1⅛	51 900	509	51.1
32	1¼	67 800	665	66.7
35	1⅜	81 100	795	79.8
36	–	85 900	842	84.5
38	1½	95 600	938	94.1
40	–	106 000	1 040	104

To obtain calculated aggregate breaking loads multiply the figures given in columns 3, 4 and 5 by 1.200.
1 kgf/mm² = 100 kgf/cm².

NON-ROTATING ROPES

These ropes (Fig. 1.18) warrant special consideration, handling and care since they are much more easily damaged in service than any other type of rope.

The non-rotating characteristic is secured by building into the rope two layers of strands, one having right lay and the other left lay. The tendency of one layer of strands to rotate in one direction is counteracted by the tendency of the other layer of strands to rotate in the opposite direction. (Fig. 1.19)

They require very careful handling prior to, during and after installation if good service is to be obtained. The manufacturer's lay length should not be disturbed by twisting or turning the ropes when installing, otherwise core slippage may occur. As non-rotating ropes, by their design, are counter-balanced, they have

TABLE 1.6

ROUND STRAND EQUAL LAY 6 × 26 to 6 × 41 SERIES INDEPENDENT WIRE ROPE CORE – BS.302

- 6 × 26 (10/5 and 5/5/1)   9 to 40 mm diameter
- 6 × 31 (12/6 and 6/6/1)   11 to 40 mm diameter
- 6 × 36 (14/7 and 7/7/1)   13 to 52 mm diameter
- 6 × 41 (16/8 and 8/8/1)   16 to 60 mm diameter

1	2	3	4	5
Nominal diameter	Approx. equivalent diameter	Minimum breaking load at 180 kgf/mm² (115 tonf/in²)		
mm	in	kgf	kN	tonf
9	3/8	5 190	50.9	5.11
10	–	6 420	63.0	6.32
11	7/16	7 770	76.2	7.65
12	–	9 230	90.5	9.08
13	1/2	10 800	106	10.6
14	9/16	12 500	123	12.3
16	5/8	16 400	161	16.1
18	11/16	20 700	203	20.4
19	3/4	23 100	227	22.7
20	13/16	25 700	252	25.3
22	7/8	31 000	304	30.5
24	15/16	36 900	362	36.3
26	1	43 300	425	42.6
28	1 1/8	50 300	493	49.5
32	1 1/4	65 700	645	64.7
35	1 3/8	78 500	770	77.3
36	–	83 200	816	81.9
38	1 1/2	92 600	908	91.1
40	–	103 000	1 010	101
44	1 3/4	124 000	1 220	122
48	–	148 000	1 450	146
52	2	174 000	1 710	171
54	–	187 000	1 830	184
56	2 1/4	201 000	1 970	198
60	–	231 000	2 270	227

To obtain calculated aggregate breaking loads multiply the figures given in columns 3, 4 and 5 by 1.285.
1 kgf/mm² = 100 kgf/cm².

TABLE 1.7

ROUND STRAND EQUAL LAY 6 × 19 (9/9/1) INDEPENDENT WIRE ROPE CORE – BS.302

1	2	3	4	5
Nominal diameter	Approx. equivalent diameter	Minimum breaking load at 180 kgf/mm² (115 tonf/in²)		
mm	in	kgf	kN	tonf
8	5/16	4 110	40.3	4.05
9	3/8	5 220	51.2	5.14
10	–	6 440	63.2	6.34
11	7/16	7 790	76.4	7.67
12	–	9 270	90.9	9.12
13	1/2	10 900	107	10.7
14	9/16	12 600	124	12.4
16	5/8	16 500	162	16.2
18	11/16	20 800	204	20.5
19	3/4	23 200	228	22.8
20	13/16	25 700	252	25.3
22	7/8	31 100	305	30.6
24	15/16	37 000	363	36.4
26	1	43 500	427	42.8
28	1 1/8	50 400	494	49.6
32	1 1/4	65 900	646	64.9
35	1 3/8	78 800	773	77.6
36	–	83 400	818	82.1
38	1 1/2	92 900	911	91.4

To obtain calculated aggregate breaking loads multiply the figures given in columns 3, 4 and 5 by 1.270.
1 kgf/mm² = 100 kgf/cm².

no tendency to twist or spin either way. However, they are so pliable that 'turn', in or out, can easily be imparted. If turns are put into these ropes then the outer strands become shorter in their initial length, the core slips along and may protrude from the rope, the outer strands become overloaded as the core then is not taking its share of the load. When turns are taken out of these ropes core slip again occurs, the outer strands bulge out, or 'bird-cage' (Fig. 1.52), and the inner layers or

WIRE ROPES

TABLE 1.8

**MULTI-STRAND 17 × 7 or 18 × 7
FIBRE CORE – BS.302**

NOTE: These ropes are generally referred to as 'non-rotating' but it should be recognised that this is a relative term.

1	2	3	4	5
Nominal diameter	Approx. equivalent diameter	Minimum breaking load at 180 kgf/mm² (115 tonf/in²)		
mm	in	kgf	kN	tonf
8	5/16	3 670	36.0	3.61
9	3/8	4 640	45.5	4.57
10	–	5 730	56.2	5.64
11	7/16	6 940	68.1	6.83
12	–	8 260	81.0	8.13
13	1/2	9 690	95.1	9.54
14	9/16	11 200	110	11.0
16	5/8	14 700	144	14.5
18	11/16	18 600	182	18.3
19	3/4	20 700	203	20.4
20	13/16	22 900	225	22.5
22	7/8	27 800	273	27.4
24	15/16	33 000	324	32.5
26	1	38 800	381	38.2

To obtain calculated aggregate breaking loads multiply the figures given in columns 3, 4 and 5 by 1.283.
1 kgf/mm² = 100 kgf/cm².

core will become overloaded by taking all of the load.

Extreme care must be taken, when securing a plain end, to ensure that the entire cross-section of the rope is firmly secured. Otherwise core slippage can occur (Fig. 1.20) and the wire core may draw (or pull) inside the rope's end and the surplus core protrude elsewhere through the outer layer. The outer layer of the strands will become overloaded if the core does not take the load.

When a non-rotating rope is to be cut, secure seizings of annealed wire should be placed tightly on both sides of the point where the cut is to be made, otherwise the balance of equal tension in all of the strands placed in the rope by the manufacturer may be disturbed by strand or core slippage. Seizings for these ropes should be put on a short distance apart, and the length should be at least equal to 1½ times or twice the diameter of the rope. There should be three on each side of the cut for ropes up to 26 mm (1 in) diameter, and four on each side when over 26 mm (1 in) in diameter.

Core slippage may also occur when bending the rope around a thimble prior to splicing; the outer strands may bird-cage or open up. For this reason thimbles of larger size than those specified for similar size ropes should be used. The rope should also be tightly served for the length of the rope around the thimble, plus twice the length for seizing. The seizing should be put on the two parts of rope at the throat of the thimble immediately after securing the thimble into the served eye of rope. The length of the throat seizing should be equal in length to the circumference of the rope. (Fig. 1.21)

Because of the small radius of wedge sockets and the possibility of severe core slippage, wedge socket anchorages should not be used on non-rotating ropes unless extreme care is used in their installation and the rope end is very tightly seized.

Where swaged end fittings are used, because of the possibility of core slippage, it is preferable to press the swaging in the hydraulic press first and after applying the annealed wire seizings previously mentioned, cut off the rope after swaging is completed. This may leave a protrusion of rope beyond the swaged fitting but this is much more preferable to core slippage.

To secure optimum non-rotating characteristics the manufacturers of non-rotating rope may recommend that they be used with a factor of safety of approximately 8 or 10:1. Some limited field experience has shown, however, that these ropes perform satisfactorily and safely on both single and multi-part lines (but not jib hoist reeving) on cranes at factors of safety of 3.5:1 provided that they are not abused and are given extremely careful examinations.

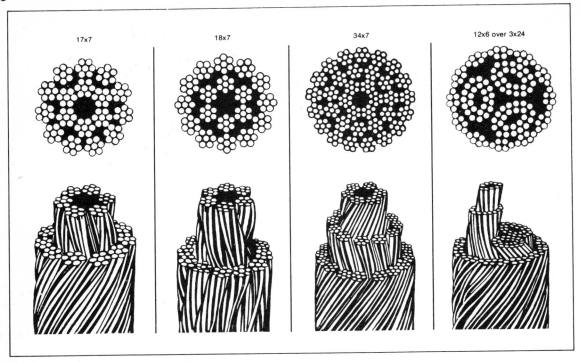

Fig. 1.18 Typical multi-strand, non-rotating constructions

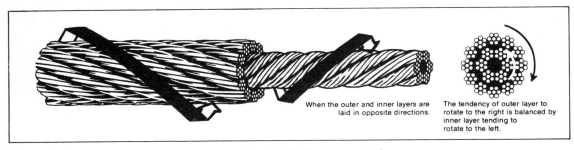

Fig. 1.19 How the anti-rotational characteristic is built into non-rotating rope

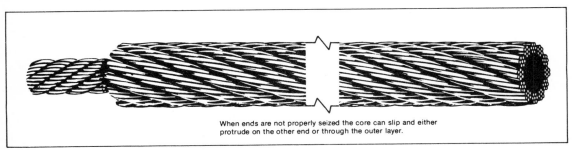

Fig. 1.20 Core slippage in a non-rotating rope

WIRE ROPES

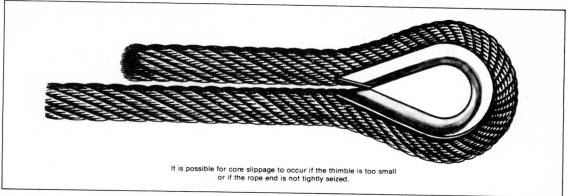

It is possible for core slippage to occur if the thimble is too small or if the rope end is not tightly seized.

Fig. 1.21 Core slippage in a non-rotating rope

INSTALLATION

Removing from shipping reel or coil: When removing the rope from the shipping reel or coil, the reel or coil must rotate as the rope unwinds. Any attempt to remove a rope from a stationary reel or coil will almost always result in a kinked rope that is ruined beyond repair at that point.

One of the three following methods (Fig. 1.22) should be used:

- Mount the reel on a horizontal shaft supported at each end by jacks or blocks. Grasp the free end of the rope and walk away from the reel which will rotate as the rope unwinds. A piece of planking should be jammed against one of the reel flanges to act as a brake to keep the rope tight and prevent the reel from overwinding.
- With the reel upright on its flanges, secure one end of the rope and roll the reel away from it. When it is necessary to lay a rope out on the floor, ensure that the surface is clean.
- Place the reel on its side, with one flange on a turntable. As the rope is pulled off, the reel turns on the turntable.

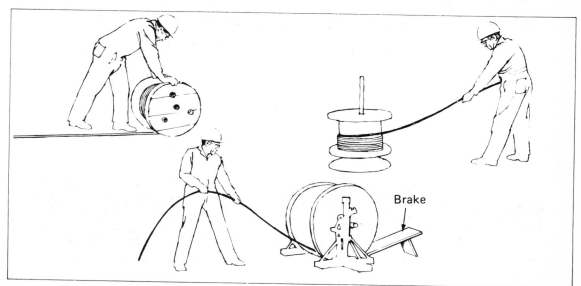

Fig. 1.22 Correct methods of removing rope from a reel

WIRE ROPES

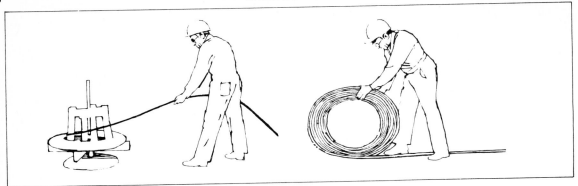

Fig. 1.23 Correct methods of removing rope from a coil

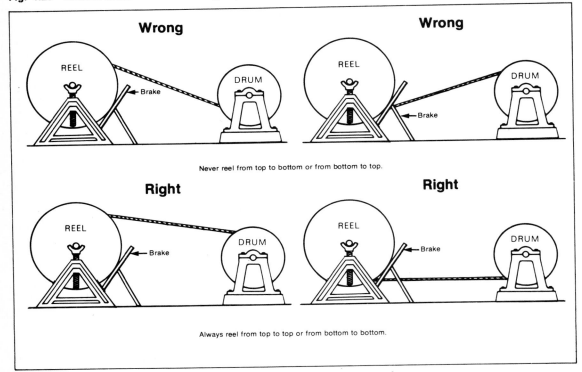

Fig. 1.24 Methods of winding rope from reel to drum or drum to reel

When unwinding a coiled rope (Fig. 1.23) use either of the two following methods:

- Secure the free end of the rope and roll the coil along the ground like a hoop. Ensure that all rope remaining in the coil is held together so that no tight coils or kinks occur. Make certain that the floor is clean.
- Lay the coil on a turntable and pull off the free end as the coil and turntable revolve. The turntable should have a centre core of approximately the same diameter as the eye in the coil. Make certain that the rope turns do not flip up and fall over the centre core.

Winding onto drum or reel: When winding a rope onto a crane drum or onto another reel, make certain that it bends in the same direction (Fig. 1.24). Re-reel from the top of one reel to the top of another, or from the bottom of one reel to the bottom of another. Follow the same procedure when winding onto crane drums. It is also necessary to apply a tensioning load to the rope to achieve good spooling.

A simple brake, such as a plank bearing against the reel flanges, will provide ample rope tension throughout the winding.

Cutting wire rope: Before cutting any wire rope (including preformed rope) apply the correct number of seizings, as per Table 1.10, on each side of the proposed cut, then cut the rope.

Locating drum anchorage point: Locate the rope anchorage point on the crane drum in relation to the direction of lay of the rope, as follows:

— Find out if the rope is to be overwound or underwound on the drum. (Fig. 1.25)

If you are installing a *right* lay rope, visualise yourself as being behind the drum; make a fist with your *right* hand to simulate the drum and extend your forefinger to simulate the rope. For overwound rope keep the back of your fist up (see how your extended finger is simulating the rope coming off the top of the drum). For underwound rope turn your palm up and your finger simulates the rope coming off the bottom of the drum. Your extended finger always points to the flange to which the rope must be secured. For *left* lay ropes follow exactly the same procedure using your *left* hand. (Fig. 1.26)

Winding onto flat-faced drums: On a flat-faced drum, it is most important that the rope should wind in a straight helix at its proper angle. This can be assisted by means of a steel starting piece cut in a taper to fill the space between the first turn and the flange. (Fig. 1.27)

Precautions should also be taken at the point where the rope enters the face of the drum at the anchorage to avoid a knuckle forming in the rope. This is particularly important for multi-layer winding where the overwinding turn would cause hammering and rapid fatigue in the knuckle section.

Turn first layers tight and true: When installing new rope it is important that all turns of the first layer on the drum be tight and true. Open or wavy winding will result in serious damage to multiple windings. Adjacent turns should be tapped against each other with a soft metal or wooden mallet, but not so that the strands interlock. (Fig. 1.28)

Rope length: Ensure that the correct length of rope is fitted. This is particularly important as it may be necessary to fit specific lengths of rope for particular applications and rope reeving combinations. Too short a rope could result in the rope completely paying out and the rope anchorage on the drum having to carry the full load. On the other hand, too long a rope could exceed the drum's spooling capacity, causing it to ride over the flanges with resulting severe damage, crushing or complete failure. It is good practice to check the rope length whenever the structure is changed, rope is changed or reeving altered.

Number of rope layers on drum: As far as is practical, the drum should accommodate all the rope in one smooth layer. More than three layers of rope result in reduced rope life and may cause crushing and severe abrasion of the rope on the bottom layer as well as at the end of any layer where pinching occurs.

Multi-layer winding: On drums with multi-layer winding, difficulty may be experienced where, as the rope rises from the first to the second layer, wedging action and severe abrasion take place. This can be overcome by the use of a tapered steel lifter to raise the rope from the first to the second layer. (Fig. 1.29)

Tight winding: The whole rope should be wound on the drum tightly and correctly as this will facilitate good winding during its full life. A poor start may mean poor winding and short life and, therefore, the extra care required for proper installation is well justified.

Spooling capacity: With the hook or load at its lowest point there should be at least two or three full turns of rope on the drum and the spooling capacity must not be exceeded when the hook is at its highest point. The flanges on grooved drums must project either twice the rope diameter or 50 mm (2 in) beyond the last layer of rope, whichever of the two is the greater. The flanges on ungrooved drums must project either twice the rope diameter, or 63 mm (2½ in) whichever of the two is the greater. (Fig. 1.30)

Condition and size of sheaves and drums: Always check the condition and dimensions of the sheave grooves before the new rope is placed in service. Table 4.4 on page 00 provides the sheave groove tolerances for new rope installations where the rope is normally oversized.

WIRE ROPES

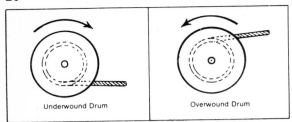

Fig. 1.25 Difference between underwound and overwound drums

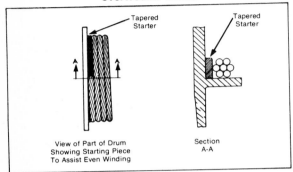

Fig. 1.27 Tapered starting piece to assist even winding

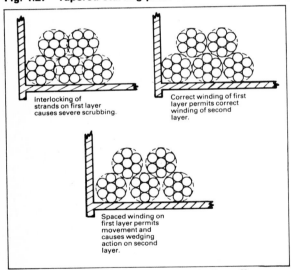

Fig. 1.28 How incorrect first layer winding affects rope life

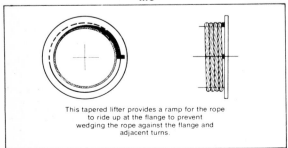

Fig. 1.29 Tapered steel lifter to assist even winding

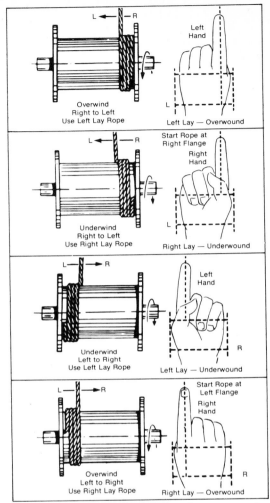

Fig. 1.26 Correct method of locating rope anchorage point on a drum

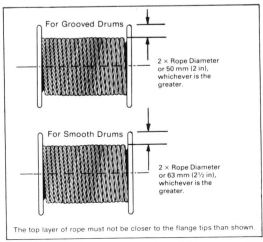

Fig. 1.30 Maximum drum capacities

WIRE ROPES

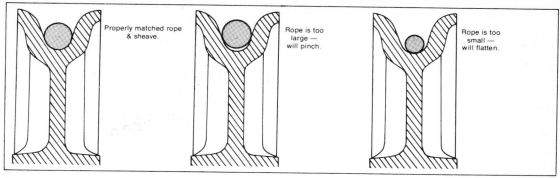

Fig. 1.31 Matching of ropes and sheaves

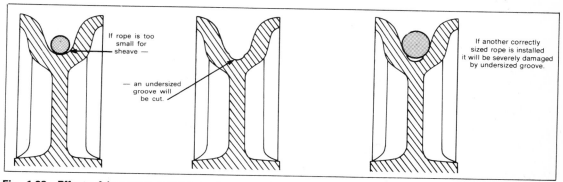

Fig. 1.32 Effect of incorrect matching of rope and sheave

When the groove diameter is slightly larger than the nominal rope diameter it provides maximum support for the rope. If the groove is too large it cannot provide adequate support and the rope may become flattened. Alternatively, too small a groove will pinch and bind the rope, causing abrasion and burning. (Fig. 1.31)

The diameter of a rope diminishes in size during lengthy service due to abrasion and loss of core support. An undersized rope may wear an undersized groove in the sheave and if a new rope is installed in the worn groove it may become wedged between the flanges. (Fig. 1.32)

To ensure a long and efficient rope life the grooves should be smoothly contoured, free of surface defects and possess rounded edges.

The tread diameter of the sheave or drum determines the bending stress in the rope and the contact pressure between it and the sheave or drum surface. Too small a sheave or drum subjects the rope to excessive bending stresses with the resulting fatigue of individual wires.

A lightly loaded rope subjected to many small-diameter bends may develop fatigue in the wires adjacent to the core of the rope — a condition almost impossible to detect.

Heavy bearing pressures introduced by a small sheave may cause wear to both sheave and rope. On applications where the major source of rope deterioration is that caused through bending or bearing pressures, large-diameter sheaves or drums should be employed to extend rope life.

Definite sheave and drum diameters cannot be specified to cover all types of operation, but Table 4.3 may be used as a guide.

Sheave alignment: After a rope change check that the sheaves are all properly aligned and running true. If they are not then considerable wear of both rope and sheaves will result with potentially disastrous consequences. A ready indication of poor alignment is rapid wear of one flange of the sheave. Poor alignment of the main sheave may also result in poor winding on the drum.

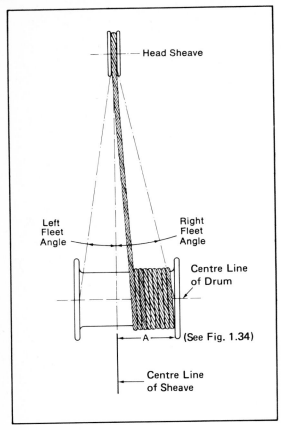

Fig. 1.33 Fleet angle relationships

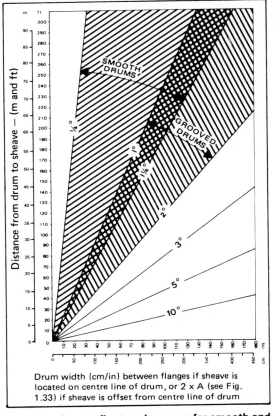

Drum width (cm/in) between flanges if sheave is located on centre line of drum, or 2 x A (see Fig. 1.33) if sheave is offset from centre line of drum

Fig. 1.34 Correct fleet angle ranges for smooth and grooved drums

Fleet angles: When a new installation is designed, the first or main sheave after the drum must be placed so as to give the right fleet angle.

The fleet angle is that angle between the line drawn from the centre of the sheave through one flange of the drum and a line through the centre of the sheave. Due to offsetting of the drum it is possible, in some installations, that the fleet angle on one side of the drum may be larger than the other; it is the larger of the two which must be considered. (Fig. 1.33)

The generally accepted standards for fleet angles are:

— Smooth drums — ¼°–1¼°
— Grooved drums — 1°–2°

BS.302 advises 1½° for smooth and 2½° maximum for grooved drums. BS.1757 and 2799 advise 1-in-12 or 4.75° maximum for all drums.

The fleet angle relationships are shown graphically in Fig. 1.34. The shaded areas are the proper range of fleet angles for both grooved and smooth drums.

If the fleet angle is larger than the recommended limits, the rope may rub badly against the flanges of the groove in the sheave or be subjected to crushing and abrasion on the drum. Too small a fleet angle will cause the rope to pile up against the flange head and damage both rope and equipment. (Fig. 1.35)

Severe scuffing of grooved drums can result when the rope wears against the groove sides. This action also bruises and crushes the rope.

On the other hand the angle should not be too small (e.g. a long distance between the drum and the sheave) as considerable vibration or dancing of the rope results with subsequent deterioration of the rope due to hammering of the rope as it winds on the drum. The installation of intermediate idler sheaves will eliminate this effect. (Fig. 1.36)

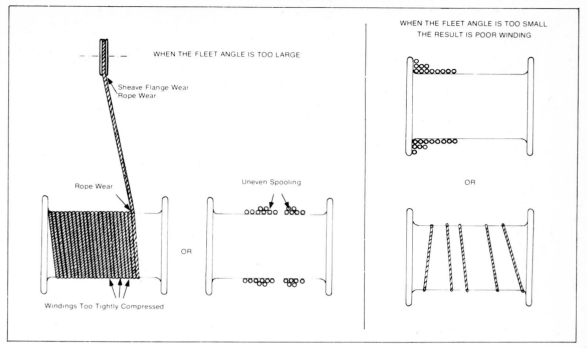

Fig. 1.35 Effects on winding of incorrect fleet angles

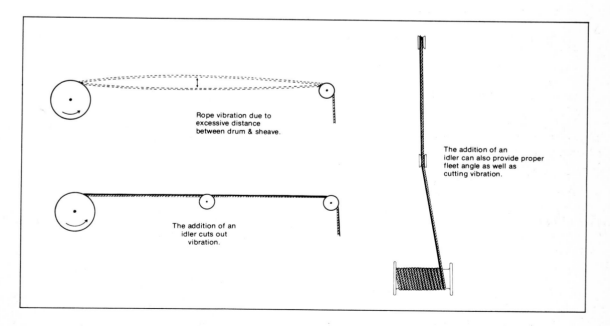

Fig. 1.36 Effect of excessive distance between drum and first sheave

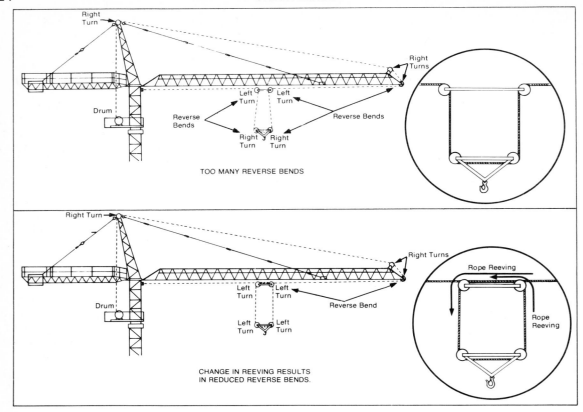

Fig. 1.37 Elimination of reverse bends in wire rope

Reverse bends: Check that so far as is practicable there are no reverse bends in the rope reeving. They accelerate rope deterioration because the rope is bent first in one direction and then the other. This results in accelerated rope fatigue at a rate more than twice what it would be were there no reverse bends.

As the rope passes over a sheave it is bent, and as it leaves the sheave it is straightened. Two distinct actions causing fatigue. This is accentuated if the rope, after being bent in one direction is then straightened and again bent in an entirely opposite direction over another sheave, after which it is again straightened. It must be appreciated that this reverse bending action initiates much greater fatigue than if all bends were made in the one direction. (Fig. 1.37)

Where reverse bends are unavoidable, the sheaves causing the reverse bend should be spaced as far apart as is practicable and the most offending sheaves should be made about ⅓ to ½ larger than the others.

Even the equalising sheave on a hoist block should be kept larger than the critical diameter, for there is enough bending to cause fatigue even though this is not a true operating sheave.

Run-in period: After installing a new rope it is advisable to run it through its operating cycle several times under light load and at reduced speed. This allows the new rope to adjust itself gradually to working conditions and enables the strands to become seated, for some stretch to occur, and for the diameter to reduce slightly as the strands and core are compacted. The rope is then less liable to be damaged when full load is applied.

The initial stretching (constructional stretch) is a permanent elongation that takes place due to slight lengthening of the rope lay, compression of the core and adjustment of the wires and strands to the load. Constructional stretch – which varies with the severity of the operation – generally takes place during the

WIRE ROPES

first weeks of operation, and increases the length of the rope by approximately ½% for fibre-core ropes and ¼% for steel-core ropes.

If the new rope is being reeved by being attached to the end of the old one and pulled through the system, the joint between the ropes during the transfer should not only be strong but should also prevent the twisting of the old rope from being transmitted to the new rope.

The rope must also be kept under as much tension as possible when reeving.

INSPECTION

The single, most important operational check to be made on lifting machines and lifting tackle equipment is the rope and rigging inspection. Assurance of safety and economy in use of the equipment dictates the requirement for a programme of periodic inspections and examinations of all load-supporting wire rope and fittings. Factors such as abrasion, wear, fatigue, corrosion, kinking and incorrect reeving are often of greater significance in determining the usable life of wire rope than are strength factors based on new rope conditions.

All wire rope in continuous service should be observed during normal operation and visually inspected weekly. A complete and thorough examination of all ropes in use must be made at least once a month and all rope which has been idle for a period of a month or more should be given a thorough examination before it is put back into service. All inspections and examinations should be the responsibility of and be performed by an appointed competent person who makes a complete report of the rope's condition. (Fig. 1.38)

The number of hours per day, week, month or year during which the rope is in use is important. Where the rope is in constant use, a thorough inspection should be made regularly once a week or more often if required. A record of each rope should be kept (include date of

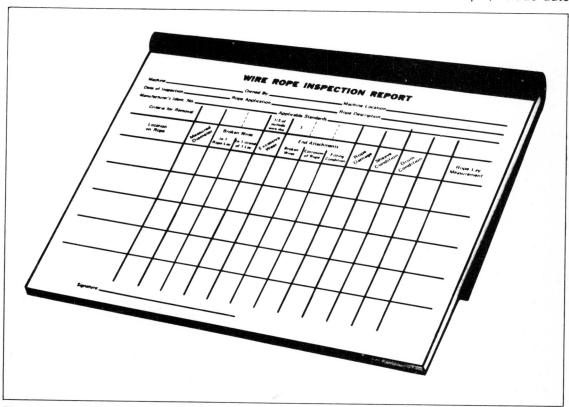

Fig. 1.38 Sample wire rope inspection log

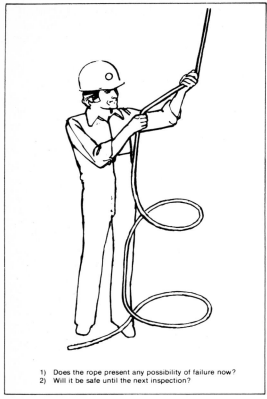

1) Does the rope present any possibility of failure now?
2) Will it be safe until the next inspection?

Fig. 1.39 Criteria for wire rope replacement

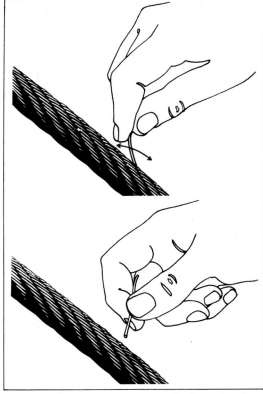

Fig. 1.40 Correct way to remove broken wires from rope

fitting, size, construction, length, defects found during inspections and examinations, and length of service).

It is good practice, where the equipment is constantly in use, to give the rope a certain length of service, several hundred hours, several weeks or months and then renew the rope regardless of its condition. This method eliminates the risk of fatigue causing rope failure.

Any deterioration, resulting in a suspected loss of original rope strength, should be carefully examined and a determination made as to whether further use of the rope would constitute a safety hazard.

The time to remove a rope from service is related to the conditions of the particular installation. These conditions include the size, nature and frequency of the lifts, when the next examination will be, what the operating and maintenance practices are and the extent of possible or probable injury to people, loss of life, material damage, etc., should the rope fail.

The user of the rope is the person most familiar with these conditions and as such should have the final responsibility of determining the maximum allowable deterioration before the rope must be removed from service.

Only by examination can it be determined whether or not the rope should be replaced (Fig. 1.39). The inspector must decide:

— If the rope's condition presents any possibility of failure, and
— If the rate of deterioration of the rope is such that it will remain in safe condition until the next scheduled examination.

When inspecting or examining the rope give every part of its length equal care as serious deterioration frequently occurs in localised positions. The estimate of the rope's condition must be made at the section showing the *maximum* deterioration.

Conditions such as the following are sufficient either to seriously question the rope's safety or to remove it from service immediately and replace it:

(1) *Broken wires:* Occasional premature wire failures may be found early in the life of almost any rope and in most cases they should not constitute a basis for rope removal provided they are at well-spaced intervals. Note the area and watch carefully for any further wire breaks. The broken wire ends should be removed as soon as possible by bending the broken ends backwards and forwards with a pair of pliers or the fingers (if possible) (Fig. 1.40). In this way the wire is more likely to break inside the rope where the ends are left tucked away between the strands where they will do no harm. Ripping the broken ends off with pliers is likely to leave jagged ends that can cut and wear unbroken wires.

The rope must be replaced if:

(a) In running ropes, there are six or more randomly-distributed broken wires in one rope lay, or three or more broken wires in one strand in one rope lay (Fig. 1.41). (A rope lay is the length along the rope in which one strand makes a complere revolution around the rope.)
(b) In pendants or standing ropes, there are three or more broken wires in one rope lay.
(c) In any rope there is one or more broken wires near an attached fitting (Fig. 1.42). Breaks occurring near attached fittings, such as sockets, are usually the result of fatigue stresses concentrated in these localised sections. Wire breaks of this type should be cause for replacement of the rope or renewal of the attachment to eliminate the locally fatigued area. About 2 m (6–8 ft) should be cropped off the rope below the socket.
(d) In running ropes there is any evidence of wire breaks in the *valleys* between strands (Fig. 1.43). Breaks occurring on crowns of outside wires indicate normal deterioration. Breaks in valleys between strands indicate an abnormal condition, possibly fatigue or breakage of other wires not readily visible. More than one of these valley breaks in one rope lay should be cause for replacement.

Note: *Construction (Lifting Operations) Regulation 34(3)* limits visible broken wire to 5% of total wires in a length of 10 rope diameters.

(2) *Worn and abraded wires:* (Fig. 1.44) Each individual wire in a rope, when new, is a complete circle in cross section. Wear, due to friction on sheaves, rollers, drums, etc., eventually causes the outer wires to become flat on the outside, reducing the circle to a segment which gradually becomes smaller as the 'flat' increases. These worn areas become void of lubrication and are characterised by their bright appearance. Close examination will reveal that the wires are much flatter in appearnace than the surrounding wires. This is part of normal service deterioration and in most installations where operating conditions are not particularly severe, relatively even abrasion will occur on the outer wires. **The rope must be replaced, however, if this wear exceeds ⅓ of the diameter of the wire.**

(3) *Reduction in rope diameter:* (Fig. 1.45) Any marked reduction in rope diameter is a critical deterioration factor. It is often due to excessive abrasion of the outside wires, loss of core support, internal or external corrosion, inner wire failures or a loosening of the rope lay. All new ropes stretch slightly and decrease in diameter after being used. This is normal, but the rope must be replaced if the diameter is reduced by more than:

– 1 mm (3/64 in) for rope diameters of up to and inluding 19 mm (¾ in)
– 1.5 mm (1/16 in) for rope diameters of 22–28 mm (⅞–1⅛ in)
– 2 mm (3/32 in) for rope diameters of 32–38 mm (1¼–1½ in)

(4) *Rope stretch:* (Fig. 1.46) Severe stretch or elongation of rope is also a deterioration factor. All steel ropes will stretch during their initial periods of use. This is known as 'constructional stretch' and it is caused by the tightening of the wires and strands into their respective cores. An approximate elongation of 150 mm (6 in) per 30 m (100 ft) of rope can be expected in a six-stranded rope and approximately

WIRE ROPES

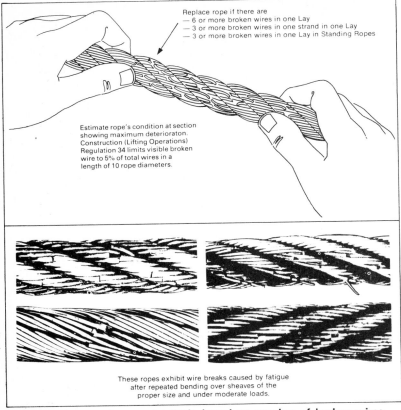

Fig. 1.41 Rope replacement criteria based on number of broken wires

Fig. 1.42 Wires broken near fittings

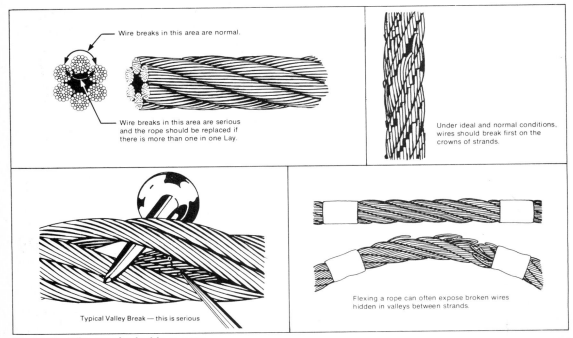

Fig. 1.43 Wire breaks inside a rope

WIRE ROPES

Normal Surface Wear

Severe wear in Langs Lay, caused by abrasion at cross-over points on multi-layer drum windings.

Severe wear and protrusion of core caused by high bearing pressure on drum and sheaves.

Section Through Worn Section

Enlarged View of Single Strand

When the surface wires are worn by $1/3$ or more of their diameter the rope must be replaced.

Normal surface wear of $1/3$ outer wire dia. on Langs Lay rope.

Normal surface wear of $1/3$ outer wire dia. on Regular Lay rope.

Rope strand indicating the result of severe surface wear and inter-strand nicking, caused by excessive rope loading.

Fig. 1.44 Worn and abraded ropes

230 mm (9 in) to 255 mm (10 in) in an eight-stranded rope. **Excessive stretch beyond this should be cause for replacement.** Watch for a lengthening of rope lay or a reduction in rope diameter. These are the signs of severe stretch, which is generally caused by overloading or a loss of strength as the rope approaches the end of its life cycle.

(5) *Corrosion:* (Fig. 1.47) This can be infinitely more dangerous than wear. More wires are affected by corrosion and visual field inspections do not give even an approximate idea of the quality of a corroded rope. This is because corrosion frequently develops inside the rope before any evidence is visible on rope surface. If corrosion is detected by the characteristic discoloration of the wires or, in particular, if pitting is observed then consideration must be given to replacing the rope. Noticeable rusting and the development of broken wires in the vicinity of attachments is also cause for replacement. If the corrosion occurs at the base of the socket then it must be cut off.

(6) *Insufficient lubrication:* Check whether the lubrication is sufficient; a rope is usually lubricated internally by the saturated fibre core. However, this either dries out or is heated or squeezed out. Examine the grooves between the strands. Where these are filled with hard-packed grease or dirt the lubricant cannot penetrate to prevent internal friction. The rope should be scrubbed and a fairly viscous warm oil applied to allow it to enter the inside of rope, cool, and remain on the rope without being thrown off when revolving around the sheaves.

(7) *Damaged or inadequate splices:* (Fig. 1.48) All splices must be closely examined for worn and broken wires, pinched or jammed strands, loose strands, cracked fittings, tucks drawing out, corrosion, loose servings, etc. If any of these conditions are evident then that section of the rope must be scrapped and a new splice made.

(8) *Corroded, cracked, bent, worn and incorrectly applied end connections:* (Fig. 1.49) If any of these conditions exist, replace the fitting. Examine all thimbles closely for wear in the crown, for evidence of the throat biting into the rope and for distortion or closure of the thimble (evidence of overloading).

(9) *Crushed, flattened or jammed strands:* (Fig. 1.50) Replace the rope as these conditions are dangerous due to severe wire deformation. These conditions can occur when there are multiple layers on drums. Ropes with a large number of wires (i.e. 6 × 37) should never be used on multi-winds. The wires are small for flexibility, but suffer badly from crushing. A larger-wired rope (i.e. 6 × 19) should be used but sheave and drum sizes must be increased accordingly to reduce fatigue due to bending. Alternatively, independent steel wire cored ropes should be used to prevent crushing. These defects can also occur if the hoist rope becomes slack and cross-coiled on the drum or trapped in the machinery. No further operations should be carried out until the rope has been paid out, examined by a competent person for possible damage, and correctly re-spooled.

(10) *High stranding and unlaying:* (Fig. 1.51) Replace the rope or renew the end connection to reset the rope lay. In cases such as this, excessive wear and crushing take place and the other strands become overloaded.

(11) *Bird-caging:* (Fig. 1.52) Replace the rope of the affected section of the rope.

(12) *Kinks:* (Fig. 1.53) Replace the rope or the affected section of the rope. They are usually caused by faulty handling or reeving. The strands become dog-legged and where running on sheaves are subject to excessive wear at the kink.

(13) *Bulges in rope:* Replace the rope, particularly if it is of a non-rotating construction. This is indicative of core slippage or turns being put into or taken out of the rope.

(14) *Gaps or excessive clearance between strands:* Replace the rope.

(15) *Core protrusion:* (Fig. 1.54) Replace the rope.

(16) *Unbalanced severely worn areas:* (Fig. 1.55) Remove the affected section.

(17) *Heat damage, torch burns, electric arc strikes:* Remove either the affected section or the whole rope.

When inspecting or examining a rope remember that the rope speed has a bearing on

WIRE ROPES

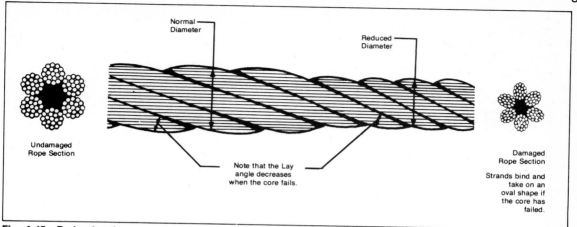

Fig. 1.45 Reduction in rope diameter

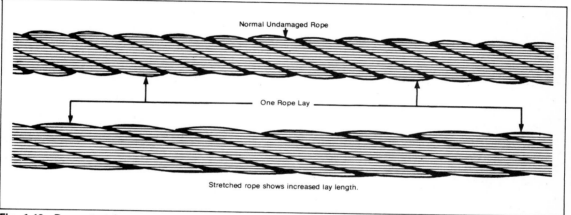

Fig. 1.46 Rope stretch

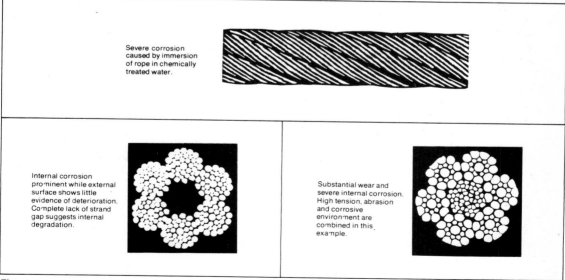

Fig. 1.47 Rope corrosion

WIRE ROPES

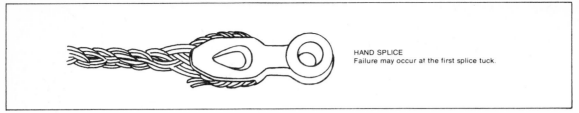

Fig. 1.48 Damaged splice

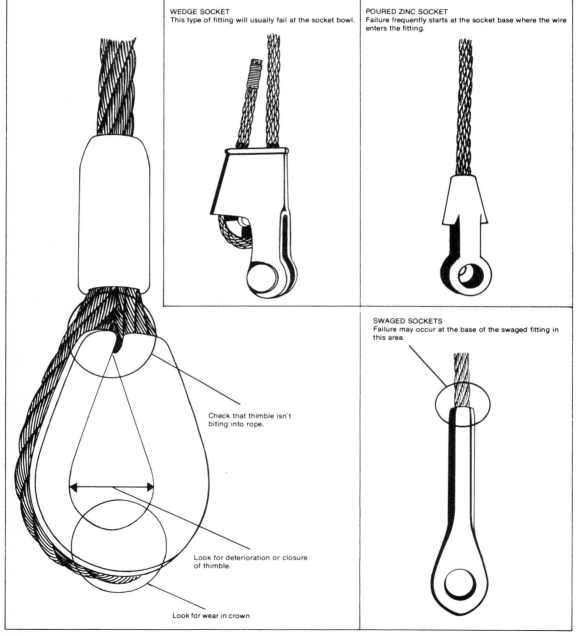

Fig. 1.49 Damaged end fittings

WIRE ROPES

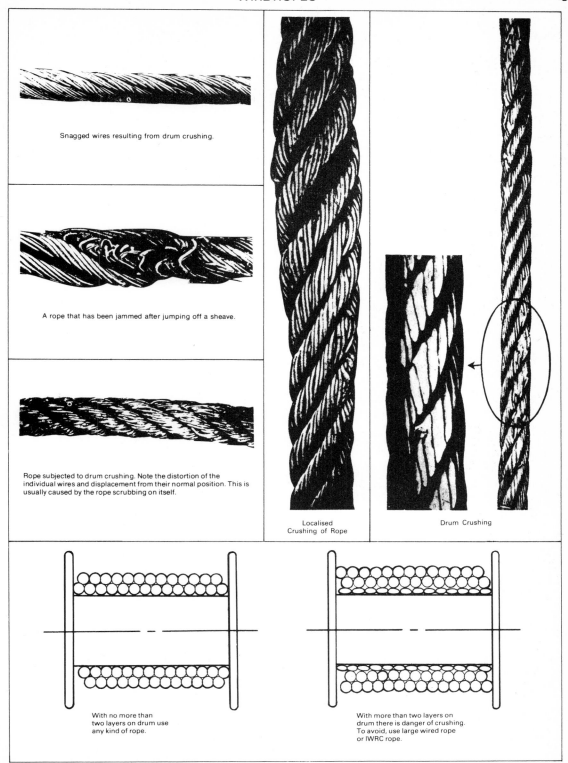

Fig. 1.50 Crushed jammed and flattened strands

the life of a rope. The life expectancy of a high-speed rope, due to increased impact effects at sheaves and drums, friction and abrasion, is less than a slow-speed rope. Due consideration must be given to this aspect and the inspections made accordingly.

If the rope being examined comprises multi-layers of strands, open it up and examine the inner strands. (Fig. 1.56)

Particular attention should be paid to those areas close to the terminal fittings. Where multi-layer drums are used, examine not only that part of the rope which is in constant use, but also the rope which may remain spooled and inoperative on the drum. Examine all ropes, including standing ropes, for possible defects caused by corrosion, abrasive dust, and erection and dismantling procedures.

Permanent damage and deformation such as kinks, crushed, flattened, and distorted strands and unbalanced wear locations are the conditions which make a rope extremely susceptible to failure.

When replacing a rope, make certain that the replacement rope is of the correct size and construction. One of the most useful aids in selecting the correct wire rope is to examine the worn ropes that are to be replaced. Knowing what caused the deterioration of the old rope may indicate a direction for change. Table 1.9 and Fig. 1.57 may aid in this determination.

The ways in which individual wires of a rope break can also give an indication of the probable cause of failure can be seen in Fig. 1.58.

Like a chain, the weakest part of a rope determines the strength of the rope. This weak point may be where the rope is badly worn, where it has a number of broken wires, where the stresses are unequal in the strands, where the local stresses due to bending the rope over a sheave are excessive, or where the rope is attached to the equipment upon which it is used. It is only by inspection and examination that potential hazards can be detected and accidents be avoided.

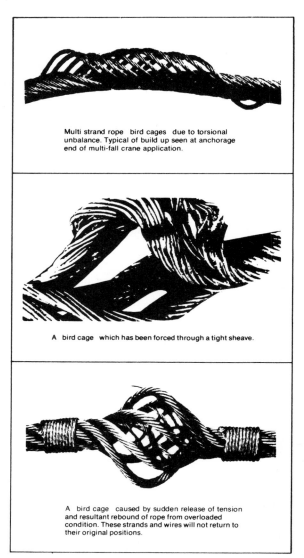

Multi strand rope bird cages due to torsional unbalance. Typical of build up seen at anchorage end of multi-fall crane application.

A bird cage which has been forced through a tight sheave.

A bird cage caused by sudden release of tension and resultant rebound of rope from overloaded condition. These strands and wires will not return to their original positions.

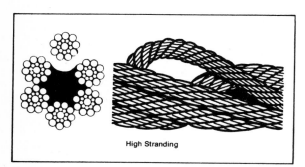

High Stranding

Fig. 1.51 High stranding

Fig. 1.52 'Bird cages'

WIRE ROPES

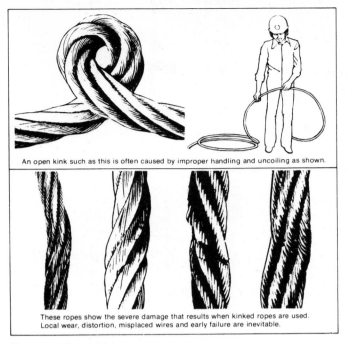

An open kink such as this is often caused by improper handling and uncoiling as shown.

These ropes show the severe damage that results when kinked ropes are used. Local wear, distortion, misplaced wires and early failure are inevitable.

Fig. 1.53 Rope kinks

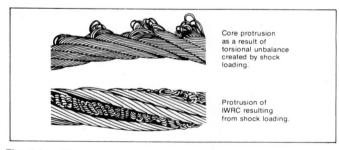

Core protrusion as a result of torsional unbalance created by shock loading.

Protrusion of IWRC resulting from shock loading.

Fig. 1.54 Core protrusion

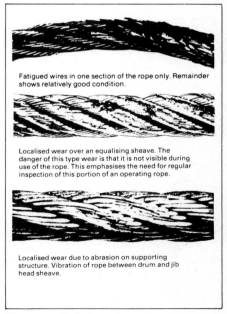

Fatigued wires in one section of the rope only. Remainder shows relatively good condition.

Localised wear over an equalising sheave. The danger of this type wear is that it is not visible during use of the rope. This emphasises the need for regular inspection of this portion of an operating rope.

Localised wear due to abrasion on supporting structure. Vibration of rope between drum and jib head sheave.

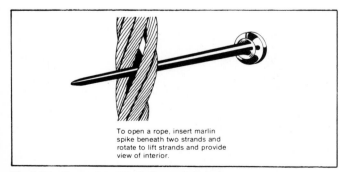

To open a rope, insert marlin spike beneath two strands and rotate to lift strands and provide view of interior.

Fig. 1.56 Correct method of opening out a rope

Fig. 1.55 Unbalanced severely-worn areas

WIRE ROPES

Narrow path of wear resulting in fatigue fractures, caused by working in a grossly oversize groove, or over small support rollers.

Break up of IWRC resulting from high stress application. Note nicking of wires in outer strands.

Two parallel paths of broken wires indicative of bending through an undersize groove in the sheave.

Wire fractures at the strand, or core interface, as distinct from crown fractures, caused by failure of core support.

An example of fatigue failure of a wire rope which has been subjected to heavy loads over small sheaves. The usual crown breaks are accompanied by breaks in the valleys of the strands, these breaks being caused by strand nicking resulting from the heavy loads.

Wire rope that shows severe wear & fatigue from operating over small sheaves with heavy load and severe abrasion

A rope failing from fatigue after bending over small sheaves.

A wire rope which has jumped a sheave. The rope itself is deformed into a 'curl' as if bent around a round shaft. Close examination of the wires show two types of breaks — normal tensile 'cup and cone' breaks and shear breaks which give the appearance of having been cut on an angle with a cold chisel.

Mechanical damage due to rope movement over sharp edge projection while under load.

Rope break due to excessive strain.

Fig. 1.57 Typical rope damage (continued opposite)

WIRE ROPES

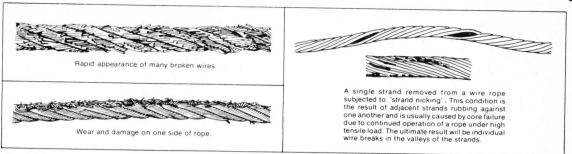

Fig. 1.57 Typical rope damage (continued)

TABLE 1.9

FAULT	POSSIBLE CAUSE	FAULT	POSSIBLE CAUSE
Accelerated wear	Severe abrasion from being dragged over the ground or obstructions. Rope wires too small for application or wrong construction or grade. Poorly aligned sheaves. Large fleet angle. Worn sheaves with incorrect groove size or shape. Sheaves, rollers and fairleads having rough wear surfaces. Stiff or seized sheave bearings. High bearing and contact pressures.	Broken wires or undue wear on one side of rope	Incorrect alignment. Damaged sheaves and drums.
		Broken wires near fittings	Rope vibration.
		Burns	Sheave groove too small. Sheaves too heavy. Sheave bearings seized. Rope dragged over obstacle.
Rapid appearance of broken wires	Rope is not flexible enough. Sheaves, rollers, drums too small in diameter. Overload and shock load. Excessive rope vibration. Rope speed too high. Kinks that have formed and been straightened out. Crushing and flattening of the rope. Reverse bends. Sheave wobble.	Rope core charred	Excessive heat.
		Corrugation and excessive wear	Rollers too soft. Sheave and drum material too soft.
		Distortion of lay	Rope incorrectly cut. Core failure. Sheave grooves too big.
		Pinching and crushing	Sheave grooves too small.
Rope broken off square	Overload, shock load. Kink. Broken or cracked sheave flange.	Rope chatters	Rollers too small.
		Rope unlays	Swivel fittings on Langs lay ropes. Rope dragging against stationary object.
Strand break	Overload, shock load. Local wear. Slack in 1 or more strands.	Crushing and nicking	Rope struck or hit during handling.
Corrosion	Inadequate lubricant. Incorrect type of lubricant. Incorrect storage. Exposure to acids or alkalis.	High stranding	Fittings incorrectly attached. Broken strand. Kinks, dog legs. Incorrect seizing.
Kinks, dog legs, distortions	Incorrect installation. Incorrect handling.	Reduction in diameter	Broken core. Overload. Corrosion. Severe wear.
Excessive wear in spots	Kinks or bends in rope due to incorrect handling in service or during installation. Vibration of rope on drums or sheaves.	Bird cage	Sudden release of load.
Crushing and flattening	Overload, shock load. Uneven spooling. Cross winding. Too much rope on drum. Loose bearing on drum. Faulty clutches. Rope dragged over obstacle.	Strand nicking	Core failure due to continued operation under high load.
		Core protrusion	Shock loading. Disturbed rope lay. Rope unlays. Load spins.
Stretch	Overload. Untwist of Langs lay ropes.		

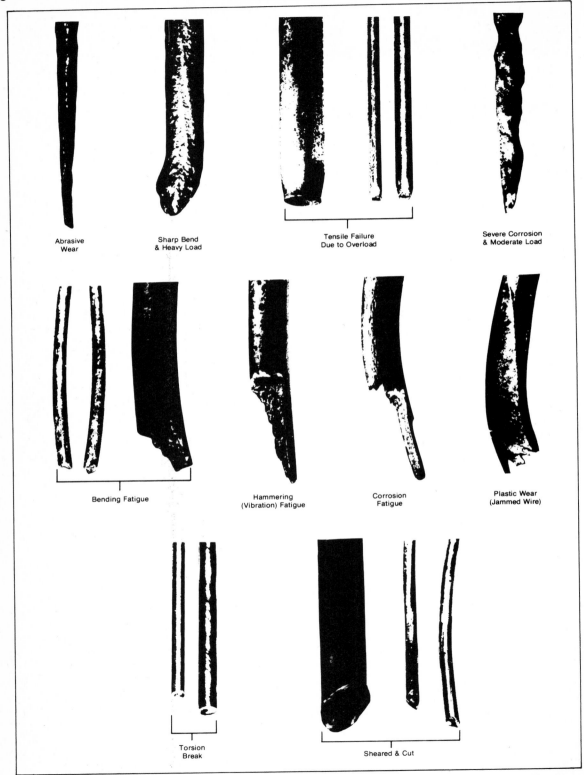

Fig. 1.58 Typical wire failures

WIRE ROPES

USE, HANDLING AND MAINTENANCE

PROCEDURES AND PRECAUTIONS

Wire ropes, like the machines and hoists on which they are used, require careful use, handling and maintenance for satisfactory performance, long life and adequate safety.
The following precautions should be observed to meet these requirements:

- Ensure that the correct rope is used.
- Inspect or examine regularly following the rope manufacturer's guidelines and the recommendations in the section on Inspection.
- Never overload.
- Minimise shock loading as overstressing of the rope will occur. In order to ensure that there is no slack in the rope at the start of loading, start the load carefully and apply the power smoothly and steadily.
- Avoid sudden loading in cold weather.
- Never use frozen ropes.
- Take special precautions and/or use a larger-sized rope whenever
  (i) the exact load is unknown,
  (ii) there is a possibility of shock loading,
  (iii) the conditions are abnormal or severe,
  (iv) there is a hazard to personnel.
- Protect rope from sharp corners or edges with padding.
- Avoid dragging the rope from under loads or over obstacles.
- Avoid dropping the rope from heights.
- Avoid rolling loads with ropes.
- Store all unused rope in a clean, dry place.
- Never use wire rope which has been cut, badly kinked or crushed.
- Prevent loops in slack lines from being pulled tight and kinking. Once a kink has been made in a wire rope the damage is permanent. A weak spot will aways remain no matter how well the kink seems to have been straightened out. If a loop forms don't pull it out, unfold it.
- Ensure that the drums and sheaves are of sufficient diameter.
- Avoid reverse bends.
- Repair or replace faulty guides and rollers.
- Ensure the sheaves are aligned and the fleet angle is correct.
- Replace sheaves having deeply worn or scored grooves, cracked or broken rims, and worn or damaged bearings.
- Repair faulty clutches.
- Check for abnormal line whip and vibration.
- Ensure that the rope spools properly on the drum.
- Never wind more than the correct amount of rope on any drum.
- Never allow the rope to cross-wind.
- Ensure rope ends are properly seized.
- Ensure that the ropes do not bind in sheaves. New wire rope requires a run-in period before operating at full load and full speed.
- Use thimbles in eye fittings at all times.
- Lubricate regularly according to the rope manufacturer's recommendations.
- Watch for local wear. Premature wear at one spot is common and can be prevented if the cause is detected. Uneven wear can be minimised by moving the rope at regular intervals so that different stretches of it are at the critical wear points. Changing layer and crossover points is merely a matter of cutting a few feet of rope from the drum end and fastening it. The cut should be long enough to move the change of layer at least one full coil from its former position and to move the crossover points one-quarter turn around the drum. Move the static section on an equaliser sheave three sheave diameter lengths away by cutting off the section on the drum end of the rope. To distribute wear due to vibration, cut off a section next to the anchorage and refasten the rope.

Even with the utmost care during installation, it is quite common to find that the dead turns on the drum become somewhat slack in operation. This slackness arises from a certain amount of stretch which occurs in a new rope under tension and periodically throughout the life of the rope from release of the load. When this slackness is noted, these turns should be rewound to tighten them. If left uncorrected, a wedging action, causing abrasion from the second layer, will occur and broken wires in the dead turns will appear.

Localised abrasion and fatigue can be dealt with without the necessity of discarding the whole rope by installing a longer rope than necessary initially and cutting at one end or the other to expose a new section of rope to the place where the deterioration occurs.

WIRE ROPES

In installations where rope deterioration is excessive at either one end or the other, the life of the rope may be extended by turning the rope end-for-end. This must be done before the deterioration becomes too severe.

On equipment having multiple falls of other than non-rotating ropes, a new rope will stretch and unlay slightly, causing turns to appear in the load block. The anchorage, if not fitted to a suitable swivel, should be disconnected, the turns removed and reconnected.

Effective maintenance of the equipment over which the ropes operate has an important bearing on rope life. Worn grooves, poor alignment of sheaves and worn parts resulting in shock loads and excessive vibration will have a deteriorating effect.

Sheaves and grooved rollers should be checked periodically for wear in the grooves which may cause pinching and abrasion of the ropes. If the groove bears the imprint of the rope it should be machined clean or replaced with a sheave of harder material. The same should be done with drums showing similar effect. Sheaves with oversized grooves do not properly support the rope and must be replaced. (Fig. 1.59)

Rollers and fairleads should be checked for worn grooves which may cut the wires. It is economical practice to offset rollers from the centre line of the rope so that they may be turned end-for-end. Rubber-faced rollers can be used to advantage in some instances.

Poor alignment of sheaves will result in wear on the rope and wear on the sheave flange. This should be corrected immediately or a distorted groove will also result.

Sheave and roller bearings should be checked for free operation. Sticking will cause unnecessary wear.

Excessive wear in sheave bearings can cause rope fatigue from the vibration.

Excessively heavy sheaves should be replaced as they tend to rotate from centrifugal force after the rope stops. This causes wear on the rope.

An excessive fleet angle will cause severe abrasion on the rope as it winds onto the drum. This condition can severely shorten rope life.

LUBRICATION

The lubrication new ropes receive during manufacture is adequate for initial storage and the early stages of the rope's working life. It must be supplemented, however, at regular intervals in accordance with the rope manufacturer's instructions.

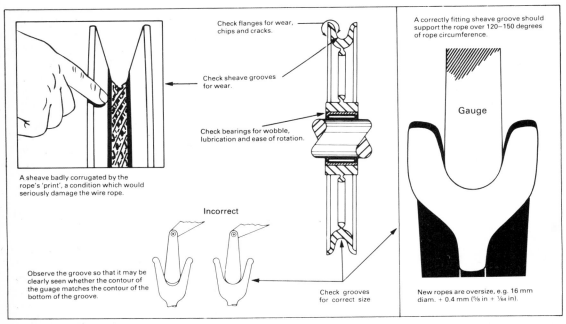

Fig. 1.59 Checking sheaves for signs of wear

WIRE ROPES

If a planned programme of regular lubrication is not carried out then the rope will deteriorate rapidly because:

- Corrosion and pitting will occur causing a loss of steel area and, therefore, a loss in strength of the rope.
- The wires will become embrittled from excessive corrosion and will break easily.
- Since each wire in the rope moves relative to the others during operation they are subject to frictional wear. Lack of lubrication will increase the wear rate causing a marked reduction in strength from loss of steel area.
- Pits also cause internal nicking of the wires which results in loss of strength.
- Ropes not in regular service or those not considered as operating ropes are vulnerable to weathering-out of the lubrication. Moisture seeps in and both core and wires deteriorate.

Lubricating a rope is as important as greasing any other piece of machinery. Consult your rope manufacturer for lubricants designed especially for the conditions inherent in an operating rope or a standing rope.

Used crankcase oil should never be used to lubricate a rope. It contains tiny metal chips which will abrade the rope, it is acidic and it has few of the characteristics that a good rope lubricant should have.

Good lubricants have the following characteristics:

- Corrosion resistance.
- Water repellant.
- Penetrating ability.
- Chemically neutral.
- High-pressure flow characteristics.
- Adhesiveness and an affinity for steel.
- Plastic coating.
- Temperature stability.

The rope must be clean and dry before the lubricant is applied because effective lubrication occurs only when the dressing comes in contact with bare metal.

If this is not done, the lubricant will fall off allowing moisture to work into the rope to cause corrosion. Excessive moisture gradually leaches out the internal lubricant.

The old lubricant can be removed by stationary or power-driven wire brushes or by compressed air jets (Fig. 1.60). It is advisable to use a light, penetrating cleaner to soften the built-up material before removing it. The lubricant supplier should be contacted regarding the proper cleaning oil to use. Do not, if at all possible, use paraffin or kerosine since too much of these will remove the internal lubricant.

In view of the small space existing between wires in the strand and strands in the rope, do not expect an externally applied lubricant to completely penetrate the rope.

The main object in external lubrication is to fill the gaps between the strands and rope so that a complete seal is provided. The lubrication should be carried out periodically to maintain this seal. The frequency required will depend on the particular installation.

Application of the lubricant may be accomplished by several methods. These should be considered with due regard to the viscosity of the compound in its state for application, length of rope involved and limitation of facilities.

Light oils may be applied by brushing or running the rope through an oil bath, by spraying, drip method and mechanical force feed. (Fig. 1.61)

For maximum penetration, the lubricant should be applied to the rope where it opens up as it travels around a sheave or winds on a drum. It is also advisable to apply the lubricant in a warm area. If the lubricant must be applied manually in an extremely cold area, however, rope lubricants are available with pour points as low as -10 to $-15°C$ (-50 to $-60°F$).

Medium-weight lubricant or heavy lubricant applied hot can be brushed on, applied by hand or by running the rope through a funnel containing the lubricant. An air blast may also be used provided that dry air only is used since any moisture would promote corrosion.

Where long lengths of rope are involved brushing and hand application are tedious. However, a thorough examination can be carried on at the same time so that two important jobs can be accomplished in one operation.

It is impossible to make general recommendations regarding the time interval that should elapse between lubrications except to reiterate that the rope should be adequately lubricated at all times, and thorough periodic examinations will indicate when this must be done.

WIRE ROPES

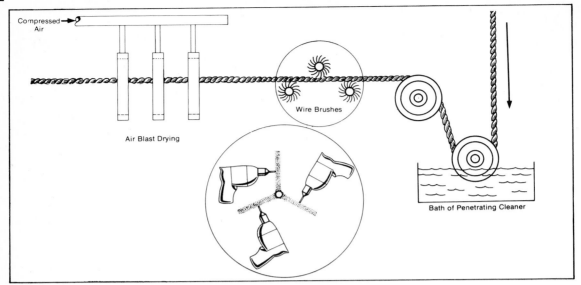

Fig. 1.60 Method of cleaning rope

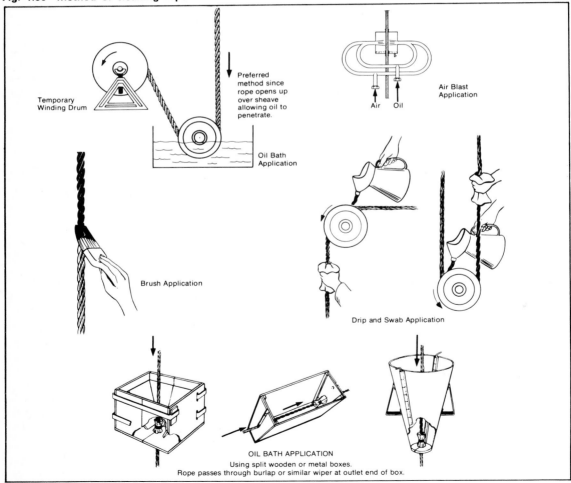

Fig. 1.61 Methods of lubricating rope

WIRE ROPES

If a wire rope is taken out of service for an appreciable length of time it should be cleaned, lubricated and stored in a dry place where it will be protected from the elements.

STORAGE

Ropes should be stored in coils or on reels in a clean, dry place indoors. If outdoor storage is necessary then the ropes must be covered to prevent the ingress of moisture or matter injurious to the rope. If lengthy storage is required it is good practice to examine the ropes periodically and lubricate them as necessary. (Fig. 1.62)

The rope must be kept away from heat and steam and must never be allowed to rest for lengthy periods of time on concrete or ash floors. The lime sulphate and ash can cause corrosion pits which act as stress raisers and eventually lead to broken wires, sometimes before the rope is put into service.

SEIZING AND CUTTING

It is most important that tight seizings of annealed wire or strand be maintained on the ends of all ropes. If this is not done the wires and strands are apt to become slack with consequent upsetting of uniformity of tensions in the rope. This results in overloading of some strands, underloading of others, birdcaging (Fig. 1.52) and breakage.

For stranded ropes there are two preferred methods of seizing:

— *For ropes larger than 26 mm (1 in) diameter,* use a soft annealed seizing wire and place one end in the valley between two strands. The long end of the wire should then be turned at right angles to the rope and wound closely and tightly back over the end of the wire and around the rope several times. Finally, the two ends of the wire should be twisted together and pulled until the seizing is tight. (Fig. 1.63)

— *For ropes smaller than 26 mm (1 in) diameter,* after winding the seizing wire on the rope, the two ends should be twisted together at approximately the centre of the seizing by alternately twisting and pulling until the seizing is tight. (Fig. 1.64)

If the rope is to be cut, seizings must be placed on both sides of the point where the cut is to be made. It can then be cut with a torch, mechanical shears, guillotines, or abrasive wheels.

For the recommended size and number of seizings see Table 1.10.

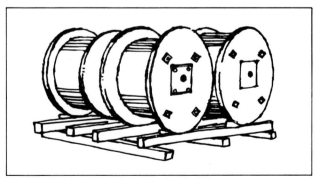

Fig. 1.62 Correct way to store wire rope reels

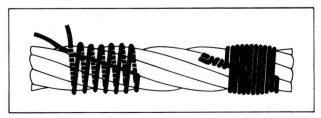

Fig. 1.63 Correct method of seizing a rope larger than 26 mm (1 in) diameter

WIRE ROPES

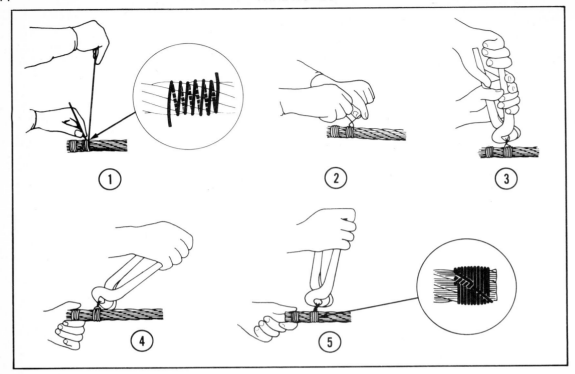

Fig. 1.64 Correct method of seizing a rope smaller than 26 mm (1 in) in diameter

TABLE 1.10

Rope diameter		MINIMUM NUMBER OF SEIZINGS				Approx. diameter of seizing			
		6- and 8-strand ropes round and flattened strand		Non-rotating ropes round and flattened strand		Stranded ropes			
mm	in	Reg. lay, fibre core	Langs lay, fibre core Reg. lay, steel core Langs lay, steel core	Groups	Seizings per group	Wire		Strand	
						mm	in	mm	in
2 and below	3/32 and below	2	2	2	3	0.5	0.020	–	–
3–6	1/8 –1/4	2	3	2	3	0.6	0.024	–	–
8–13	5/16–1/2	3	4	2	3	0.8	0.032	–	–
14–22	9/16–7/8	3	4	2	3	1.0	0.040	1.6	1/16
24–38	15/16–1 1/2	3	4	3	3	2.0	0.080	2.4	3/32
40–52	1 5/8 –2	4	5	3	3	2.6	0.104	3.2	1/8
over 52	over 2	4	5	3	3	3.3	0.128	4.0	5/32

SPLICING

The splicing of wire rope has an important bearing on the safety of the rope as insecure splices will fail if the ropes are subjected to shock loads. Also, incorrectly-made splices that disturb the construction of the rope will damage the sheaves and drums which will, in turn, damage other rope sections.

Rope splicing is a skill that requires training and proper tools to be done safely and effectively. Refer to the wire rope manufacturer's handbooks and manuals for the proper splicing types, methods and procedures to use.

END FITTINGS AND CONNECTIONS

End attachments of wire rope installations are of the greatest importance to safety and it is important to know that many wire rope attachments, even when properly made and installed, develop less than the full strength of the rope. Not only must you know what type to use and how to install them correctly, but you must know their safe working loads as compared to that of the rope. Safety requires knowing how to:

– Select the correct fittings and connections.
– Install them correctly.
– Evaluate their safe load capacity.

It is extremely important that all fittings be of adequate strength for the application. Whenever possible, 'load rated' (SWL stamped on the fitting) fittings should be used. For overhead lifting the *only* type recommended is the *forged fitting* of weldless construction.

Many types are available and the choice depends on various factors of the installation.

Zinc and swaged socket attachments are used on the more permanent types of installation such as pendants. They are suitable only for standing ropes subject to little or no movement, as rope movements tend to crack the ropes at entry to sockets. Rapid hoisting, road travelling and impact loads, acceleration or braking loads always set up severe vibration in the rope directly above the point of the load attachment. There is no way to prevent rope fatigue and eventual wire breaks from occurring at the point where the rope enters the sockets. Fig. 1.65 shows the relative fatigue life of various end fittings.

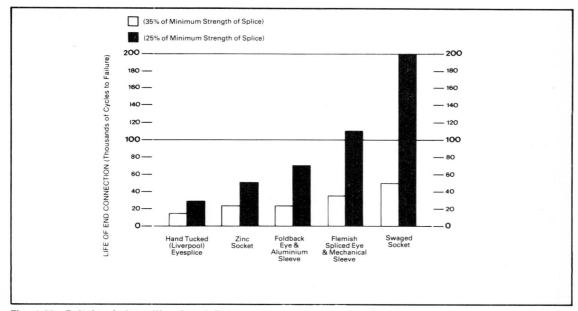

Fig. 1.65 Relative fatigue life of end fittings

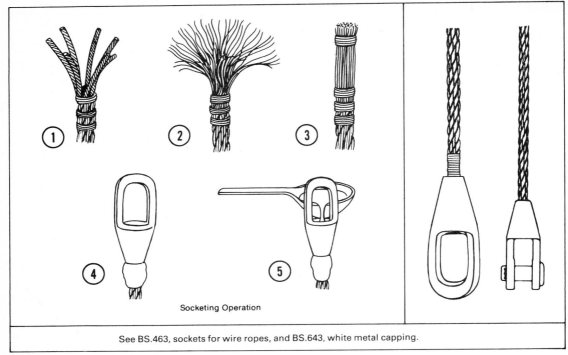

Fig. 1.66 Zinc (spelter) sockets

Zinc (spelter) sockets (Fig. 1.66) are efficient and permanent terminal attachments for wire rope. They are the most reliable of all terminal fittings and when correctly attached, these standard drop-forged sockets are 100% efficient. They are recommended for all standing ropes and whenever service conditions are severe. Only trained and fully qualified personnel should be permitted to make these connections since it is a skill that requires good facilities and a thorough understanding of the manufacturer's instructions.

Swaged sockets (Fig. 1.67) also make efficient and permanent terminal attachments for wire rope. They are made by compressing a steel sleeve over the rope with a hydraulic press. Correctly made they provide an efficiency of 100%.

Careful examination of the wires leading into these socketed terminals is most important because the strength of this section (between 1.5 and 3 m (5–10 ft) above the load attachment) is bound to be greatly reduced because of fatigue within the wires. On visual inspection, one broken wire is sufficient to cause it to be condemned. It is, therefore, advisable to cut this damaged section off periodically even though it may look sound. About ⅓ of the estimated life of the rope is a good interval.

When inspecting any existing socketed connection, examine the rope closely for corrosion at the socket base. This corroded wire is very susceptible to fatigue.

Cappel sockets (Fig. 1.68) when properly installed and frequently inspected, also give 100% efficiency. Their efficiency depends entirely upon the wedges being kept tight. Since the wedges are interlocked and cannot move independently of each other, the unit affords a holding power equal to the strength of any type of wire rope. They also afford the facility of frequent inspection of the whole of that part of the rope upon which the grip is exerted.

Wedge sockets (Fig. 1.69) are among the simplest devices for anchoring a wire rope for any purpose. They are intended for on-the-job attachment and for quick rope replacement. Principal advantages are simplicity, ease and speed of applying and detaching. They are also used where conditions are such that spliced eyes could not be reeved and would have to be made after the rope was in place.

The efficiency of a wedge socket is low, however; only 70% of the strength of the rope. Care must be taken that moving loads do not

WIRE ROPES

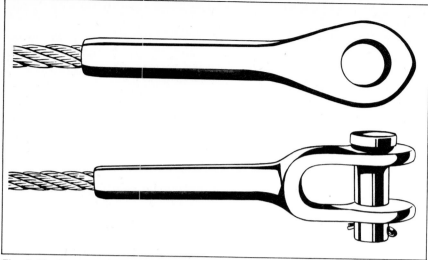

Fig. 1.67 Swaged sockets

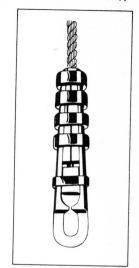

Fig. 1.68 Cappell socket

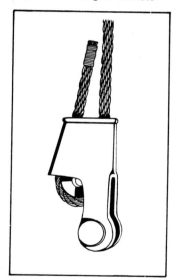

Fig. 1.69 Wedge socket

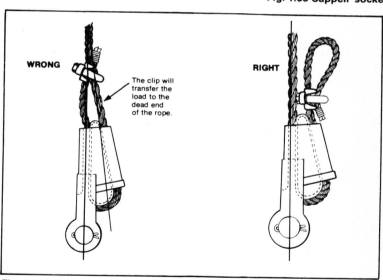

Fig. 1.70 Correct method of securing dead end of rope when using a wedge socket

force the wedges out and accidental slackening of the cable does not release the socket. To prevent this they should be positive locking. Also, the dead or short end of the cable should either have a clip attached to it, or be looped back and secured to itself by a clip. The loop thus formed must not be allowed to enter the wedge. Do not attach the dead end of the cable to the live side of the line with the clip as this will seriously weaken the attachment. The clip will ultimately take the load and could deform and break the cable. (Fig. 1.70)

Wedge socket anchorages should not be used on non-rotating ropes (because of the small radius of wedge sockets and the possibility of severe core slippage) unless extreme care is used in their installation and the rope end is very tightly seized.

When using wedge sockets it is good practice to start out with a longer rope than is required so that the socket can be renewed periodically without having to splice the rope. The wedge can be punched out of the socket, the bad wire cut off and re-fitted in the socket.

WIRE ROPES

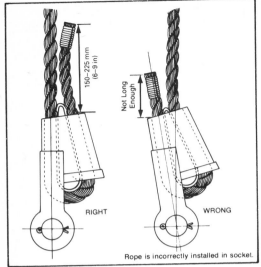

Fig. 1.71 Fitting a wedge socket on a rope

Fig. 1.72 Reason why all eyes should be fitted with thimbles

Fig. 1.73 Flemish eye and serving

Fig. 1.74 Flemish eye and pressed metal sleeve

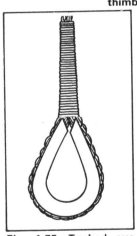

Fig. 1.75 Tucked eye and serving

Fig. 1.76 Tucked eye and pressed metal sleeve

Fig. 1.77 Fold-back eye and pressed metal sleeve

WIRE ROPES

The rope end should always protrude at least 150–225 mm (6–9 in) beyond the socket. (Fig. 1.71)

It is very important to ensure that the loaded part of the rope is not kinked where it leaves the wedge. The sockets must be installed so that the load line is in a staight line pull with the eye of the socket.

Spliced eyes in various forms are also frequently used as wire rope end attachments.

With the exception of some slings, all spliced eyes must incorporate rope thimbles to maintain rope strength and reduce wear. If a thimble is not used on a spliced eye the efficiency of the connection can be reduced by as much as 10% because the rope flattens under load. (Fig. 1.72)

Because of the many forms of eye splices, their similarity and the fact that it is very difficult to determine what kind of splice you are working with when it is covered with serving wire or pressed metal sleeves, a dangerous confusion can be created. This confusion is dangerous since there are great differences in the efficiencies of eye splices but little difference in appearance.

UK Docks Regulations 1934 (SRO 1934 No. 279) in Regulation 20 specify a minimum standard for wire rope splices.

Every wire rope manufacturer attaches a different name to his particular type of eye splice but they are usually variations of the three basic types:

Group I:
 – Flemish eye (rolled eye)
 – Flemish eye + serving (Fig. 1.73)
 – Flemish eye + pressed metal sleeve (Fig. 1.74)

Group II:
 – Tucked eye (hand-spliced eye or Liverpool splice)
 – Tucked eye + serving (Fig. 1.75)
 – Tucked eye + pressed metal sleeve (Fig 1.76)

Group III:
 – Fold-back eye + pressed metal sleeve (Fig. 1.77)

A comparison of Figs. 1.73–1.76 shows that it is almost impossible to determine what kind of splice you are working with when it is covered by serving wire or pressed metal sleeves. All personnel must be told exactly what type the company uses and all splices must be identical on all slings.

The best and most secure splices are those from Group I and the best of all and the one recommended for all lifting equipment use is the Flemish eye + pressed metal sleeve. It is an excellent attachment. When properly made it develops almost 100% of the catalogue breaking strength. The strand ends of the spliced eye are secured against the live portion of the rope by means of a steel or aluminium sleeve set in place under pressure. Of the various types of mechanical splices this is the most dependable. The strength of the splice is not wholly dependent on the sleeve since it has the hand splicing to fall back on should the sleeve fail and the tapered end of the sleeve allows the splice to pass over obstructions without snagging.

The splices from **Group II** (with the exception of the tucked eye + pressed metal sleeve which is almost 100% efficient) are less secure than those from Group I. The tucked eye, unlike the Flemish eye, can develop only 70% of the strength of the rope and tends to come free as the rope unwinds. As the ropes untwist, the tucks in the eye begin to pop free and on short lengths of cable or sling one or two turns will seriously weaken the eye and consequently they should not be used for lifting loads. This, of course, refers to the splice connection only. The addition of wire seizings protects the splice and the slinger's hands, but adds no strength.

All Group II eye splices should have at least five tucks and the completed splice should be carefully and tightly wrapped with a wire serving to cover the whole of the splice. The efficiencies of the tucked or hand-spliced eyes (without pressed metal sleeves) are as given in Table 1.11.

TABLE 1.11

Rope diameter		Efficiency
mm	in	
6 and below	¼ and below	95%
8–19	5/16–¾	90%
22–26	⅞ –1	85%
28–38	1⅛ –1½	80%
40–52	1⅝ –2	75%
54 and over	2⅛ and over	70%

The **Group III** eye splices are made by bending the rope to the eye dimension required against the live portion of the rope by means of a steel or aluminium sleeve set in place under pressure. Some manufacturers

WIRE ROPES

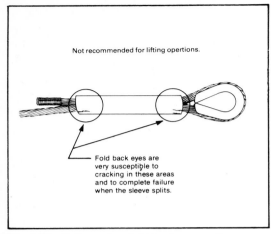

Fig. 1.78 Failure of a fold-back eye splice

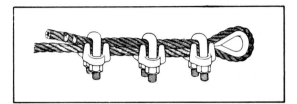

Fig. 1.80 Clamp-and-thimble connection

Fig. 1.81 Cable clip connection

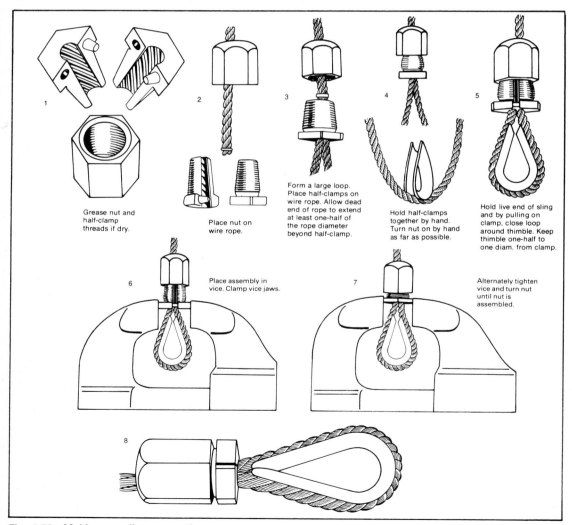

Fig. 1.79 Making a collet connection

WIRE ROPES

unlay or separate the dead end strands and re-lay them into the grooves between the strands around the body of the live side of the rope. To hold these strands in place, steel ferrules are pressed onto the splice by a hydraulic press.

It is strongly recommended that these splices never be used for overhead lifting operations since incorrect swaging or split sleeves will result in complete failure without warning. (Fig. 1.78)

It is very easy to confuse the excellent mechanically-spliced eye with this fold-back eye so be certain that all spliced eyes on the job are of one type or another, preferably of the mechanically-spliced type.

Collet connections (Fig. 1.79) are similar to the fold-back eye except that instead of the sleeve being swaged on, it takes the form of a split collet, sleeve and nut arrangement. The nut is placed on the rope, a loop of the eye size required is made. The collets are placed on the rope and the dead end of the rope is extended at least one-half of the rope diameter beyond them. The collets are held together by hand and the nut turned on as far as possible by hand. The loop is closed around the thimble by holding the live end of the rope and pulling on the clamp. The thimble is kept one-half to one rope diameter from the clamp. The assembly is then placed in a vice and the nut is torqued.

Clamp-and-thimble connections (Fig. 1.80) combine both the clamp and the thimble in one unit, and are capable of developing approximately 80% of the wire rope strength.

Cable clips (Fig. 1.81) The most common method used to make an eye or attach a wire rope to a piece of equipment is with cable or Crosby clips of the U-bolt-and-saddle type or of the double integral saddle-and-bolt type (known as 'safety' or 'fist grip').

These terminations have the advantage of allowing thorough examination and ease of field installation. When applied with proper care, thimbles, and according to the following tables, clipped eye terminations will develop approximately 80% of the rope strength. All clips must be of drop-forged steel; malleable iron clips must never be used.

U-bolt clips must have the U-bolt section on the dead or short end of the rope and the saddle on the live or long end of the rope. The wrong application (U-bolt on live instead of dead end) of even one clip can reduce the efficiency of the connection to 40%. (Fig. 1.82)

Never use fewer than the number of clips recommended in Table 1.12. Turn back the correct amount of rope for dead ending to permit proper spacing of the clips. Always use *new* clips; re-used clips will not develop the correct efficiency. It is equally important always to use a thimble to prevent the rope from wearing in the eye and to provide a safer connection.

TABLE 1.12

INSTALLATION OF WIRE ROPE CLIPS

Rope diameter		Minimum no. of clips	Amount of rope turn back from thimble		Torque unlubricated bolts	
mm	in		mm	in	Nm	lbf ft
3	1/8	2	83	3 1/4	–	–
5	3/16	2	95	3 3/4	–	–
6	1/4	2	120	4 3/4	20	15
8	5/16	2	140	5 1/2	40	30
9	3/8	2	165	6 1/2	61	45
11	7/16	2	178	7	88	65
13	1/2	3	290	11 1/2	88	65
14	9/16	3	305	12	130	95
16	5/8	3	305	12	130	95
19	3/4	4	457	18	175	130
22	7/8	4	482	19	305	225
26	1	5	660	26	305	225
28	1 1/8	5	865	34	305	225
32	1 1/4	6	940	37	488	360
35	1 3/8	7	1120	44	488	360
38	1 1/2	7	1220	48	488	360
40	1 5/8	7	1295	51	583	430
44	1 3/4	7	1345	53	800	590
52	2	8	1800	71	1015	750
56	2 1/4	8	1850	73	1015	750
64	2 1/2	9	2135	84	1015	750
70	2 3/4	10	2540	100	1015	750
76	3	10	2690	106	1625	1200

Apply the first clip one base width from the dead end of the wire rope (Fig. 1.83). Tighten the nuts. Apply the second clip adjacent to the thimble, but don't tighten the nuts. Apply all the other clips, leaving equal space between each. For maximum holding power they should be installed six-to-seven diameters apart. Take up the rope slack by applying tension to the eye and cable and tighten all the nuts evenly on all the clips to the recommended torque.

After the rope has been in operation for an hour or so, all nuts on the clip bolts will have to be retightened, and they should be checked for

WIRE ROPES

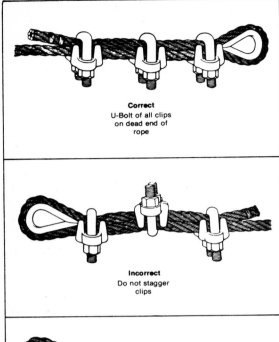

Correct
U-Bolt of all clips on dead end of rope

Incorrect
Do not stagger clips

Incorrect
U-Bolt of all clips on live end of rope

Fig. 1.82 Correct and incorrect methods of using cable clips

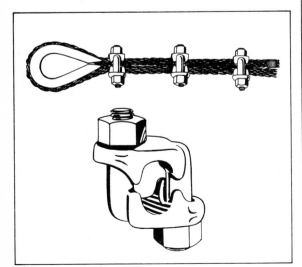

Fig. 1.84 Double saddle (or fist grip) clips

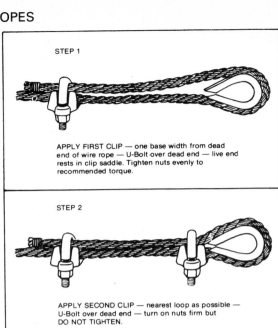

STEP 1

APPLY FIRST CLIP — one base width from dead end of wire rope — U-Bolt over dead end — live end rests in clip saddle. Tighten nuts evenly to recommended torque.

STEP 2

APPLY SECOND CLIP — nearest loop as possible — U-Bolt over dead end — turn on nuts firm but DO NOT TIGHTEN.

STEP 3

ALL OTHER CLIPS — Space equally between first two.

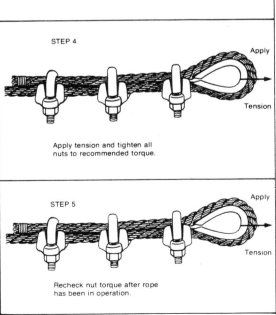

STEP 4

Apply tension and tighten all nuts to recommended torque.

STEP 5

Recheck nut torque after rope has been in operation.

Fig. 1.83 Correct method of fitting a cable clip

WIRE ROPES

tightness at frequent intervals thereafter. This is necessary because the rope will stretch slightly, causing a reduction in diameter which will slacken the clips.

Double saddle or fist grip clips (Fig. 1.84) are preferable to the U-bolt clips. It is impossible to install them incorrectly and they cause less damage to the rope. Less turn back is required and fewer double saddle clips than U-bolt clips are required in some rope sizes, as indicated in Table 1.13.

Still greater efficiency can be obtained by the use of long, double-base clamps extending at least six rope diameters in length; these apply a greater clamping force on the rope without damaging it, and the rope life and safety increase accordingly. (Fig. 1.85)

No type of clip should be used to connect directly two straight lengths of rope. If two ropes have to be joined use the clips with a thimble to form an eye at the end of each length and connect the eyes together. (Fig. 1.86)

TABLE 1.13

INSTALLATION OF DOUBLE SADDLE CLIPS						
Rope diameter		Minimum no. of clips	Amount of rope turn back		Torque unlubricated bolts	
mm	in		mm	in	Nm	lbf ft
5	3/16	2	102	4	41	30
6	1/4	2	102	4	41	30
8	5/16	2	127	5	41	30
9	3/8	2	140	5½	61	45
11	7/16	2	165	6½	88	65
13	1/2	3	280	11	88	65
14	9/16	3	324	12¾	175	130
16	5/8	3	343	13½	175	130
19	3/4	3	406	16	305	225
22	7/8	4	660	26	305	225
26	1	5	940	37	305	225
28	1 1/8	5	1040	41	488	360
32	1 1/4	6	1400	55	488	360
35	1 3/8	6	1575	62	678	500
38	1 1/2	6	1675	66	678	500

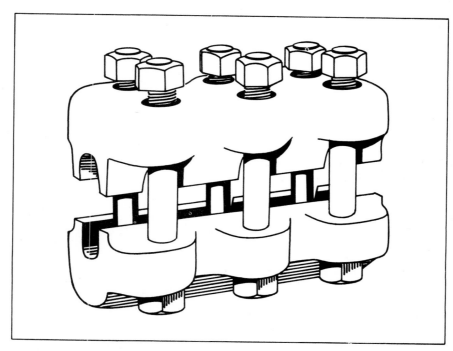

Fig. 1.85 Double base clamp

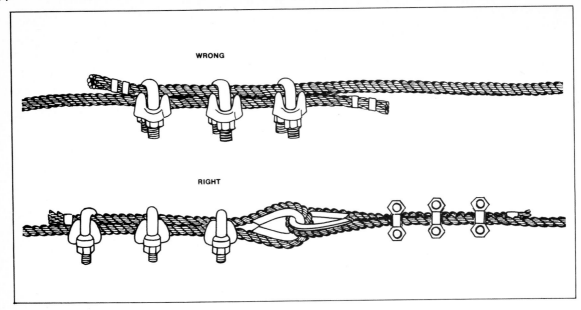

Fig. 1.86 Joining wire ropes

CHAPTER 2

Fibre rope

TYPES AND CHARACTERISTICS

Fibre ropes are made from either natural or synthetic fibres. Natural fibres come from plants and include manila, sisal and hemp, and the synthetic fibres include nylon, polypropylene, and the polyesters. The strength of these ropes depends on their size, the fibre used, and the type of stranding.

NATURAL FIBRE ROPES

The two most commonly-used natural-fibre ropes are manila and sisal, but the only type suitable for construction lifting equipment is grade 1 manila. It is strong and durable and must be used where dependability, ability to stand up under severe use, and weathering resistance are required. Sisal fibres are less durable and lower in strength and are generally made up into ropes used where the requirements are less demanding and cost is a major factor. Sisal rope is whitish, rather coarse and feels harsh to the touch.

The grades of manila rope most commonly used are:

(1) *Yacht-rope:* The highest quality manila of very fine appearance. Costly, but used on special jobs where appearance is an important factor.
(2) *Bolt-rope:* High-grade manila, about 10–15% stronger than grade 1 manila.
(3) *Grade I:* The standard high-grade rope used for important lifting operations. Usually identified by a trade marker.
(4) *Grade II:* Has about the same initial strength as grade I, but loses its strength *more rapidly* in service.
(5) *Grade III:* Has about the same initial strength as grade I, but loses its strength *very rapidly* in service.
(6) *Hardware store rope:* A very poor grade of manila, of low strength and short life.

See also BS. 2052 which deals with ropes made from manila, sisal, hemp, cotton and coir.

In purchasing manila rope for lifting purposes, the specifications should require that it be made of first-grade manila fibres, and the order should be placed with a reputable manufacturer. Most rope manufacturers place a marker of some kind in their first-grade rope for identification. Some have one or more yarns coloured, or they may have a coloured string in one strand, or they may have a tightly twisted paper ribbon with their trade mark on it in one of the strends. Rope without such a marker should not be used for lifting purposes.

In ordering new rope the purchaser should specify the type of construction best suited to his requirements. Manila rope usually comprises three strands, each containing a number of threads. The rope itself is twisted in a right-hand direction, each strand in a left-hand direction, and the individual threads in a right-hand direction. This reversal of twist in the strand gives the rope a balance or set and

FIBRE ROPE

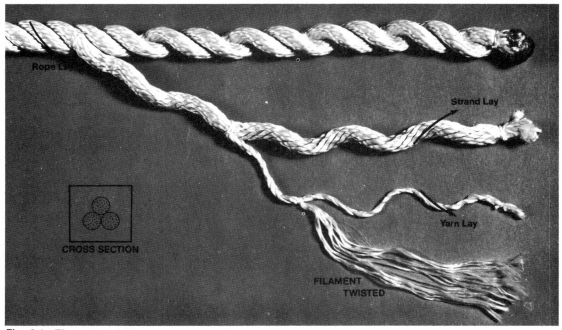

Fig. 2.1 Three-strand fibre rope construction

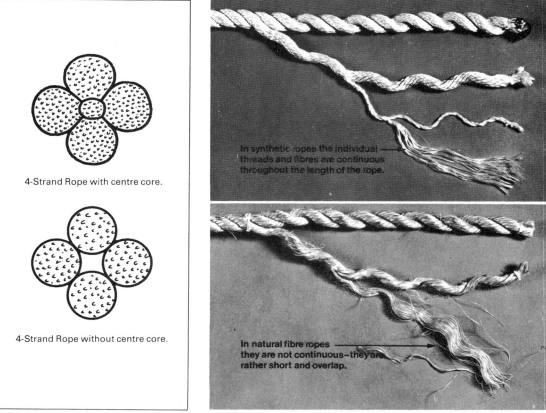

Fig. 2.2 Four-strand rope

Fig. 2.3 Difference between natural and synthetic fibre ropes

eliminates the tendency towards unwinding. (Fig. 2.1 shows a typical rope construction.)

For work where abrasion is a factor, a rope having four strands laid around a small rope core is recommended. This rope is more nearly round in cross section and is used quite extensively for power-operated lifting. Its strength, however, is slightly less than 3-strand rope. (Fig. 2.2)

In making a rope, the yarns can be formed into the strands and the strands can be laid into the rope either tightly or loosely, making what is known respectively as 'hard-laid' or 'soft-laid' rope. Hard-laid rope is stiffer and more resistant to abrasion, whereas soft-laid rope is limp but stronger. A medium lay is generally recommended.

Manila rope should be light yellow in colour, with a silvery lustre, and should have a smooth waxy surface. The grades have the following relative shades:

- *Grade I* – very light in colour.
- *Grade II* – slightly darker.
- *Grade III* – considerably darker.
- *Hardware store rope* – still darker, and the short fibres cause many ends to protrude from the strands.
- *Yacht and bolt-rope* – very light in colour.

All manila ropes are subject to rot, even though they are chemically-treated to repel moisture and resist mildew and dry rot. The best way to slow down the rotting process is to maintain the rope efficiently. They will withstand long periods of use under wet, dirty and rot-producing conditions, provided they are cleaned and dried at frequent intervals and stored so that the air can circulate around them. It is important to understand that manila ropes shrink when wet and often are of lower strength in the wet condition due to uneven distortion of the rope structure. The best cleaning method involves washing the rope in cold water and hanging it in loose folds or coils over pegs so that the air can dry it. Since mildew and rot are the most common causes of natural fibre rope deterioration, it is good economic sense to maintain the ropes in a clean and dry state. Slings and safety lines should receive special attention in this regard.

Prolonged exposure to sunlight will also cause manila ropes to deteriorate; where possible they should be either covered or shaded.

General purpose manila rope, if dried out after use and properly stored, needs no added lubricant. If the rope becomes stiff, however, a thin coat of warm lubricating oil applied with a paint brush will make the rope pliable again.

SYNTHETC FIBRE ROPES

Synthetic fibre ropes, particularly nylon and polypropylene, have rapidly gained acceptance and, to a great degree, are replacing manila rope. These ropes have individual fibres running their entire length, rather than short, overlapped fibres as in natural fibre ropes. Consequently, they have greater strength. (Fig. 2.3)

Synthetic fibre ropes are generally impervious to rot, mildew and fungus and have good resistance to chemicals. Nylon absorbs very little moisture and polypropylene none; consequently the ropes do not stiffen when wet, do not freeze and have good dielectric properties when clean and dry.

Synthetic ropes are stronger than natural fibre ropes, nylon being about 2½ times the strength of manila. They are also lighter, are easier to handle and have excellent impact, fatigue and abrasion resistance. They will also outwear manila ropes by 4 or 5:1.

They are liable to melt at high temperatures, however, and should not be used where they are likely to encounter excessive heat or where friction is sufficiently high and concentrated enough to melt the fibres. Avoid using them near welding operations.

The synthetic ropes are not generally affected by mildew or dry rot, and can withstand long periods of wetting without any noticeable loss of strength or change in appearance.

The important features of the most commonly-used synthetic fibre ropes for lifting purposes are as follows:

Nylon: This is the best-known type and it has gained the widest acceptance of the synthetic fibre ropes. It is strong, roughly 2½ times stronger than manila, has good abrasion resistance qualities and good resistance to weathering. It is usually chalky white in colour, with a smooth surface, is soft and pliant and has a feeling of elasticity to it.

FIBRE ROPE

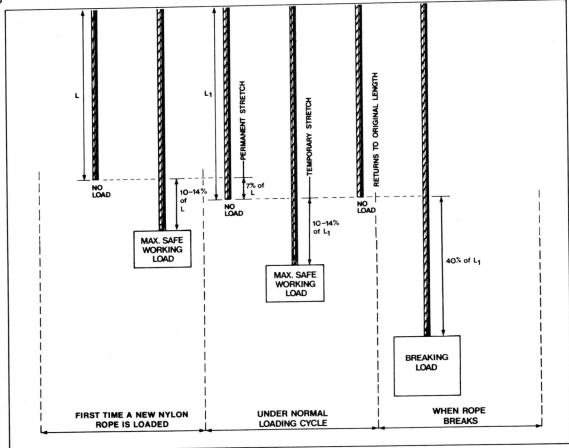

Fig. 2.4 Stretch characteristics of nylon rope

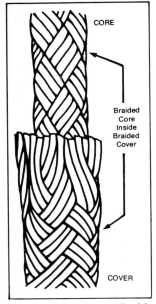

Fig. 2.5 Two-in-one braided rope

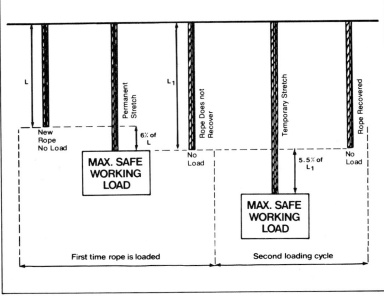

Fig. 2.6 Stretch characteristics of polyester ropes

Some properties of nylon ropes are:
- High breaking strength (wet or dry).
- Light weight per unit of strength.
- Excellent elasticity and tensile recovery.
- Superior absorption of impact and shock loads.
- Excellent flex and abrasion resistance.
- Good flexibility.
- Excellent resistance to rotting.
- Good sunlight and weather resistance.
- High melting point.

Nylon ropes are the strongest available because the filaments from which the ropes are made run continuously through the length of the rope. This also means that resistance to creep under sustained loads is very high.

Nylon rope has good resistance to abrasion because the outer fibres, if abraded, form a protective fuzz which slows down the abrasion of the rope's inner fibres.

Nylon is highly elastic and ropes are capable of withstanding severe shock loads. Because of this feature it should be used where high energy absorption is needed or where shock loading is a factor to contend with. The high degree of stretch (10–40%) is an excellent feature in some respects but a serious disadvantage in others since it can produce problems where headroom for lifting is restricted and slings have to be as short as possible.

The elongation of nylon rope (Fig. 2.4) under actual service conditions is considerably less than its ultimate breaking elongation, 10–14% under working loads and over 40% at its breaking loads. The first time a nylon rope is loaded, the rope fibres compact and the rope is permanently stretched when the load is released. The increase in length is approximately 7%. However, once this permanent stretch is in the rope it recovers completely from subsequent stretching under load. Nylon ropes should be broken-in prior to field use to allow them to stretch.

Nylon does not become progressively softer with rise in temperature but retains its physical properties almost to its melting point. Ropes can be used at temperatures up to 150°C (300°F) for long periods without serious loss of strength and at slightly higher temperatures for limited periods of time.

Nylon ropes absorb moisture and when wet lose approximately 10% of their strength. They also become very slippery. Their full strength is regained when the rope dries out.

Nylon ropes can be stored wet or dry without fear of attack by rot or mildew but good storage conditions away from heat and protected from exposure to sunlight are recommended. Unnecessary outdoor exposure to sunlight and weather should be avoided whenever possible by covering the rope.

Nylon is highly resistant to alkalis, but is rapidly attacked by most acids, paints and linseed oil, and it is recommended that all contact with chemicals be avoided. If contamination is suspected, wash the rope thoroughly in cold water and then examine the fibres carefully for evidence of weakening.

Braided Nylon: Where the highest possible strength is required it is recommended that braided nylon ropes providing a braided sheath over a braided core be used. The load is divided equally between the sheath and the core and regardless of the outside sheath damage, 50% of the rope's initial strength remains in the core. The construction provides a soft flexible rope that does not twist or kink. (Fig. 2.5)

Twisted three-strand ropes transmit a turning, twisting motion under load but braided rope with inner core and outer sheath does not have this built-in tendency to twist. They also present approximately 50% more surface area for wear and grip, are stronger size for size, are more stable, exhibit much less stretch when working, have less permanent stretch, and greater flexibility than three-strand nylon ropes.

These 2-in-1 braided ropes are available with nylon core/nylon cover to provide high strength, more elongation and excellent wearing qualities; polypropylene core/nylon cover for strength and lower stretch; and polypropylene core/polyester cover for very low stretch, high strength and good abrasion qualities.

Polyester (trade names: Dacron and Terylene): Polyester rope is nearly identical to nylon in appearance but has little or no elastic feeling. Size-for-size, they are heavier and not as strong as nylon, but are very similar in construction. Continuous filaments throughout the rope length ensure high resistance to creep under sustained loads and the low stretch properties of this material offer advantages when lifting with limited headroom. At working loads the stretch is approximately the same as manila rope. Ability to absorb shock loads is about two-thirds that of nylon, but is much greater than manila.

The first time a polyester rope is loaded it will be permanently stretched approximately 6% but future stretching under load (approximately 5.5%) is always recovered when the load is removed. (Fig. 2.6) Thus, all polyester ropes should be given a breaking-in time to allow them to stretch.

These ropes have abrasion resistance qualities similar to nylon rope.

Polyester, like nylon, does not soften progressively with rise in temperature and it can be used in similar hot conditions. The fibre is somewhat tougher than nylon, does not lose strength when wet and has a very high resistance to the degrading effects of sunlight and weather. It is immune from attack by rot and mildew, although the same general advice on good storage conditions is relevant.

Unlike nylon, polyester is resistant to acids and attack by alkalis. All chemicals should be avoided where possible. The slings should be washed frequently in cold water.

Polypropylene: Ropes made from this material and from polyethylene are nearly identical in appearance. They are available in various colours and are smooth and pliant and somewhat slippery, particularly polyethylene.

Although polypropylene is light in weight (it floats on water), this advantage has to be considered in relation to its lower strength than either nylon or polyester and accordingly it is intermediary in use between the natural fibre ropes and the more sophisticated nylon and polyester.

Stretch properties of polypropylene ropes vary with the type of construction but, generally, they stretch somewhat more than polyester. The energy absorption properties are approximately half that of nylon.

Polypropylene ropes soften progressively with rise of temperature and this plastic behaviour, combined with a relatively low melting point, make it unsuitable for use in hot conditions.

It is not as tough as either nylon or polyester, but it does not lose strength when wet. It degrades in sunlight, but it is immune to attack by rot and mildew and is highly resistant to attack by acids and alkalis, although industrial solvents have a softening effect. It is also an excellent non-conductor of electricity.

Polyethylene: Polyethylene ropes are low in strength compared with the other synthetic fibre ropes, and soften progressively with rising temperature. They are resistant to strong and weak acids and alkalis (excepting nitric acid) and are also resistant to alcohol and bleaching solutions.

COMPARISON OF ROPES

Strength: The relative strength of fibre ropes (Table 2.1) is an important factor when selecting ropes to perform a particular task.

TABLE 2.1

RELATIVE STRENGTH OF FIBRE ROPES WHEN DRY	
Type of rope	Relative capacity (related to nylon)
Nylon	100%
Polyester	87%
Polypropylene	60%
Polyethylene	52%
Manila	37%

Table 2.2 shows how the strength of most ropes changes when they are wet.

TABLE 2.2

% CHANGE IN ROPE STRENGTH WHEN WET	
Type of rope	% change in strength
Nylon	−10%
Polyester	no change
Polypropylene	+5%
Polyethylene	+5%
Manila	−5% approx.

Synthetic fibres generally absorb very little water and, accordingly, do not swell or distort the rope body. Nylon has the greatest absorption factor. This results in a loss of strength in the nylon filaments of approximately 10%. It is important to note that the full strength is recovered on drying out, and the initial strength superiority of nylon over ropes made from the other man-made fibres gives a very adequate margin to compensate for this loss.

With natural fibre rope a loss of strength will occur due to swelling of the fibres by water absorption, with a consequent distortion of the rope body. This distortion will vary within wide limits depending upon the construction of the rope and the type of fibre. Losses up to 25–30% of the dry rope strength are possible. Waterproofing of natural fibre rope can retard the absorption of water, but any process, short of completely sealing the fibres, can only delay the absorption which will take place ultimately on prolonged immersion. Natural fibre ropes (unlike nylon) do not always come back to full strength on drying.

Ability to take shock loads: One of the more important properties of fibre ropes is their ability to absorb repeated shock loads. This capacity is a function of the extensibility and elasticity of the ropes and the materials from which they are made. The ability of a rope to absorb energy is related to the following physical properties:

(a) Breaking load, weight and breaking length.
(b) Load/extension characteristics.
(c) Recovery after repeated loadings.
(d) Limiting velocity (rate of transmission of stress).

Table 2.3 gives the energy-absorption capacities of various ropes. The Table compares ropes of equal diameter and length.

TABLE 2.3

ENERGY ABSORPTION CAPACITY	
Type of rope	Relative capacity (related to nylon)
Nylon	100%
Polyester	62%
Polypropylene	40%
Steel wire rope	22%
Polyethylene	21%
Sisal	15%
Manila	13%

Note that polyethylene rope has a very low energy-absorbtion capacity. This is because it transmits stress at a slow rate. The shock is not absorbed uniformly throughout the length of the rope, but is concentrated at a local spot where reactive heat is generated. Accordingly, ropes made from polyethylene mono filaments are not recommended for applications where shock loads are anticipated.

Ability to take sustained loads: When a load or stress is first applied to any fibre rope, it will stretch it to a degree proportionate to the magnitude of the load applied. Normally this stretch is referred to as the 'strain'. It is a characteristic of fibre ropes that, should heavy loads be sustained over periods of time, further and progressive extension takes place until the ultimate extension is reached and the rope fractures. This phenomenon of creep can have an important bearing on selecting the right rope for an application where dead loads are involved.

TABLE 2.4

	INFLUENCE OF SUSTAINED LOADING		
Type of rope	Length of time at		
	75% load	50% load	25% load
Nylon	indefinitely	indefinitely	indefinitely
Polyester	indefinitely	indefinitely	indefinitely
Polypropylene	indefinitely	indefinitely	indefinitely
Polyethylene	3–6 days (varies with temperature)	2–4 weeks (varies with temperature)	indefinitely
Manila	5 minutes	3–5 hours	indefinitely
Sisal	4 minutes	45–60 minutes	indefinitely

Table 2.4 compares the performance of various ropes while loaded to 75%, 50% and 25% of their breaking strengths. The table shows the marked superiority of ropes made from materials in which the fibres are continuous throughout the length of the yarns in the ropes. Natural fibres are held together in the yarns by surface contact and twist and, when placed under load, tend to slip. As the slip progresses the rope eventually fractures. (Fig. 2.7)

Ability to take repeated loads: In normal usage, ropes are often subjected to repeated loadings. The ability of a rope to withstand heavy loadings at frequent intervals without failure, creep or loss of energy-absorption capacity is a valuable asset to the user. No fibre rope is capable of satisfying all requirements completely but knowledge of the relative responses of ropes to repeated loading is of considerable value in determining the best rope for any application.

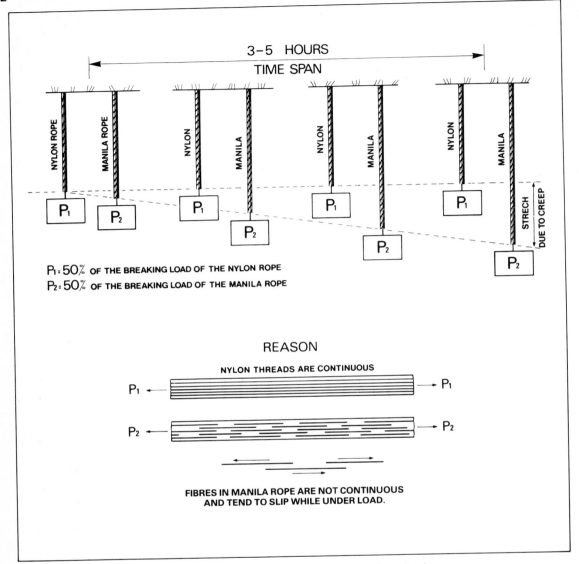

Fig. 2.7 Comparison of nylon and manila ropes under heavy sustained load

Table 2.5 rates performance under conditions of repeated loading and unloading.

Flexing endurance: The ability of a rope to take repeated bending under load is an important property. This is called 'flexing endurance' and its effect on rope performance and strength is an important factor on the length of rope life.

Table 2.6 lists the relative flexing endurance of a number of ropes in both the dry and wet states. The percentage values relate to the ratio of the number of bending cycles to failure of the rope as compared to dry nylon rope.

TABLE 2.5

\multicolumn{2}{c}{ROPE PERFORMANCE UNDER REPEATED LOADING AND UNLOADING}
Type of rope	Relative performance
Nylon Polyester Polyproplyene Polyethylene Manila Sisal	Best Decreasing Performance Worst

FIBRE ROPE

TABLE 2.6
RELATIVE FLEXING AND BENDING ENDURANCE

Type of rope	Relative resistance to failure (related to dry nylon rope)	
	Dry ropes	Wet ropes
Nylon	100.0%	55.0%
Polyester	66.0%	68.0%
Polypropylene (Monofil)	5.1%	5.4%
Polypropylene (Multifilament)	2.7%	2.6%
Sisal	1.9%	4.1%
Manila	1.4%	3.8%
Polyethylene	0.5%	2.5%

TABLE 2.7
EFFECT OF HOT SURFACES (287°F)

Type of rope	% loss of original strength
Polyester	13.2%
Nylon	16.5%
Sisal	17.0%
Manila	24.6%
Polypropylene	100.0% (distorted)
Polyethylene	100.0% (melted)

TABLE 2.8
EFFECT OF RADIATED HEAT AND HOT GASES

Type of rope	Effect
Nylon Polyester	No effect. Do not soften progressively but melt sharply at their respective melting points.
Polypropylene	Softens slowly and loses strength.
Polyethylene	Softens rapidly.
Manila Sisal	Lose strength slowly by loss of moisture, oils and lubricants and fibre embrittlement.

TABLE 2.9
STRENGTH LOSS VS. TEMPERATURE

Type of rope	Temperature (°C)				
	20	40	60	80	100
Manila	0	−3%	−8%	−20%	−30%
Nylon	0	−2%	−7%	−14%	−20%
Polypropylene	0	−7%	−18%	−30%	−40%

Effects of heat: (Tables 2.7, 2.8, 2.9) Of all the hazards to which ropes are subjected during their working lives perhaps the worst are the effects of heat since the damage done may not be apparent on normal inspection.

The principal sources of heat (with the exception of welding and cutting operations) liable to damage ropes are:

(a) Contact with hot surfaces.
(b) Radiated heat from hot gases, fires, heaters, boilers, etc.
(c) Frictional heat.

Table 2.7 shows the effect of rope contact with a hot surface (steam pipe at 140°C (287°F)) while under load for a period of 5½ hours.

If a rapidly moving loaded rope passes over a stationary rope of the same type, considerable heat develops on the stationary rope at the point of contact. This heat rapidly builds up to a temperature sufficiently high to cause melting and failure of the stationary rope. Nylon in particular is prone to breakdown of this type. This is due to its rather high co-efficient of friction and its slow rate of thermal conductivity.

Natural fibre ropes are more resistant to these conditions; they do not burn readily and breakdown is slower.

Effects of low temperatures: (Table 2.10) Ropes are often used in extremely low temperatures and it is important to know what the effects of low temperature are on the ropes.

Provided the rope is dry (i.e. containing only its normal amount of moisture) low temperatures do not reduce its tensile strength, although polypropylene does get somewhat brittle.

If a natural fibre rope is wet before cooling in cold dry air it loses considerable tensile strength. Under the same conditions, synthetic ropes lose very little strength.

FIBRE ROPE

TABLE 2.10

	EFFECT OF LOW TEMPERATURE ON ROPE STRENGTH				
	Percentage change in strength over original strength (at 21°C)				
Type of rope	Cooled (dry) in air to −10°C	Cooled (dry) in air to −10°C and warmed to 21°C	Cooled in air to −10°C after soaking in water	Cooled in air to −10°C after soaking in water and warmed to 21°C	Frozen in ice
Nylon	+ 8%	no change	+ 4%	+10%	− 9%
Polyester	+ 2%		+ 4%	+ 6%	− 2%
Polyethylene	− 4%		+ 2%	+ 2%	− 4%
Polypropylene	−15%		−12%	no change	−17%
Manila	+ 7%		−34%	−15%	−33%
Sisal	+ 9%		−35%	− 5%	−36%

In freezing conditions synthetic fibre ropes stay more flexible than natural-fibre ropes and, because of their low water-absorption properties, they handle well under sub-zero conditions.

Insulating characteristics: It is occasionally necessary to use ropes when working close to electrical power lines and a knowledge of insulating characteristics of ropes is important. (Fig. 2.8)

All ropes will conduct electricity when they are wet. However, polyester, polyethlene and polypropylene ropes all have good and consistent insulating properties under conditions of low or high humidity. Nylon is *not* recommended where insulation against high voltages is required. It absorbs moisture from the atmosphere so that, unless dried and varnished, its insulating properties vary widely.

SAFETY FACTORS

Fibre rope, like wire rope, must have a factor of safety (FoS) to account for loadings over and above the weight being lifted and for reductions in capacity due to:

— The reduced capacity of the rope below its rated strength due to ordinary usage, wear, broken fibres, broken yarns, age, variations in size and quality.
— Extra loads imposed by acceleration and inertia (starting, stopping, slewing and jerking of the load).
— Increases in line pull due to friction of the rope passing over sheaves.

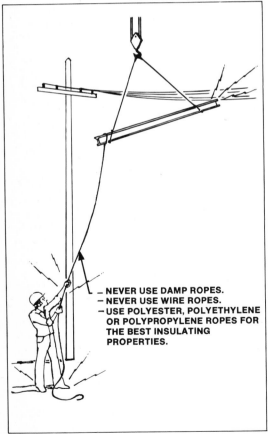

Fig. 2.8 Ropes to use and to avoid using near power lines

FIBRE ROPE

- Inaccuracies in the weight of the load.
- Reduced strength due to bending over sheaves.
- Reduced strength because of drying out, mildew and rot.
- Severe strength reductions caused by knots in the rope.
- Weakened yarns due to ground-in dirt and abrasives.

This list is not complete. It illustrates why the FoS is required and why the safe working loads (SWL) must never be exceeded. The FoS does *not* give extra usable capacity.

The FoS for fibre ropes depends on *all* the conditions of use and can vary from, say, 5:1 to 12:1 or more. The maker's or supplier's advice should be sought.

Estimate the SWL as follows:

$$SWL = \frac{\text{Minimum Breaking Load}}{\text{Factor of Safety}} = \frac{\text{Minimum Breaking Load}}{5}$$

For example, a rope with a 680 kg (1500 lb) Minimum Breaking Load has an SWL

$$\frac{680 \text{ kg (1500 lb)}}{5} = 136 \text{ kg (300 lb)}$$

SAFE WORKING LOADS (SWL)

Tables 2.11 and 2.12 giving SWL for fibre ropes are for reference purposes only. Check your rope manufacturer's ratings before determining SWL as they may differ from the Tables.

If the manufacturer quotes his rope's capacity in terms of its breaking strength, you must apply the appropriate factor of safety: in general, 12 when lifting or holding personnel and 6 for other purposes.

Table 2.11 lists the maximum SWL for common three-strand fibre ropes generally used for rigging. The figures are based on ropes without knots or hitches. Safe working loads for fibre rope slings are listed in the section on Slings.

The following rules-of-thumb work well for new ropes when load tables are not available:

MANILA ROPE

- Change the rope diameter into eighths of an inch.
- Square the numerator and multiply by 20.

Example:
(a) ½ in (13 mm) manila rope = ⁴⁄₈ in dia.
    SWL = 4 × 4 × 20 = 320 lb (145 kg)
(b) ⅝ in (16 mm) manila rope
    SWL = 5 × 5 × 20 = 500 lb (227 kg)
(c) 1 in (25 mm) manila rope = ⁸⁄₈ in dia.
    SWL = 8 × 8 × 20 = 1280 lb (580 kg)

NYLON ROPE

- Change the rope diameter into eighths of an inch.
- Square the numerator and multiply by 60.

Example:
½ in (13 mm) nylon rope = ⁴⁄₈ in dia.
SWL = 4 × 4 × 60 = 960 lb (435 kg)

POLYPROPYLENE ROPE

- Change the rope diameter into eighths of an inch.
- Square the numerator and multiply by 40.

Example:
½ in (13 mm) polypropylene rope = ⁴⁄₈ in dia.
SWL = 4 × 4 × 40 = 640 lb (260 kg)

FIBRE ROPE

TABLE 2.11

APPROXIMATE SAFE WORKING LOADS OF NEW FIBRE ROPES – KG:LB
3-strand ropes   Safety Factor = 5, as example only

Nominal diameter		Manila		Nylon		Polypropylene		Polyester		Polyethylene	
mm	in	kg	lb	kg	lb	kg	lb	kg	lb	kg	lb
5	3/16	45	100	90	200	68	150	90	200	68	150
6	1/4	55	120	135	300	113	250	135	300	113	250
8	5/16	90	200	227	500	180	400	227	500	160	350
9	3/8	122	270	320	700	227	500	320	700	227	500
13	1/2	240	530	570	1 250	376	830	545	1 200	360	800
16	5/8	400	880	910	2 000	590	1 300	860	1 900	476	1 050
19	3/4	490	1 080	1 270	2 800	770	1 700	1 090	2 400	680	1 500
22	7/8	700	1 540	1 725	3 800	1 000	2 200	1 540	3 400	950	2 100
26	1	820	1 800	2 170	4 800	1 315	2 900	1 900	4 200	1 130	2 500
28	1 1/8	1 090	2 400	2 860	6 300	1 700	3 750	2 540	5 600	1 497	3 300
32	1 1/4	1 225	2 700	3 265	7 200	1 900	4 200	2 860	6 300	1 680	3 700
38	1 1/2	1 680	3 700	4 630	10 200	2 720	6 000	4 040	8 900	2 400	5 300
40	1 5/8	2 040	4 500	5 625	12 400	3 310	7 300	4 900	10 800	2 950	6 500
44	1 3/4	2 400	5 300	6 800	15 000	3 950	8 700	5 850	12 900	3 580	7 900
52	2	2 810	6 200	8 120	17 900	4 720	10 400	6 895	15 200	4 310	9 500

See also BS.2052 (Natural Fibre Rope) and BS.4928 (Man-made Fibre Ropes)

POLYESTER ROPE

– Change the rope diameter into eighths of an inch.
– Square the numerator and multiply by 60.

Example:
1/2 in (13 mm) polyester rope = 4/8 in dia.
SWL = 4 × 4 × 60 = 960 lb (435 kg)

POLYETHYLENE ROPE

– Change the rope diameter into eighths of an inch.
– Square the numerator and multiply by 35.

Example:
1 in (25 mm) polyethylene rope = 8/8 in dia.
SWL = 8 × 8 × 35 = 2240 lb (1000 kg)

TABLE 2.12

APPROXIMATE SAFE WORKING LOADS OF NEW BRAIDED SYNTHETIC FIBRE ROPES – KG:LB
Safety Factor = 5, as example only

Nominal diameter		Nylon cover Nylon core		Nylon cover Polypropylene core		Polyester cover Polypropylene core	
mm	in	kg	lb	kg	lb	kg	lb
6	1/4	190	420	–	–	172	380
8	5/16	290	640	–	–	245	540
9	3/8	400	880	308	680	335	740
11	7/16	545	1 200	454	1 000	480	1 060
13	1/2	680	1 500	670	1 480	625	1 380
14	9/16	950	2 100	780	1 720	–	–
16	5/8	1 090	2 400	950	2 100	1 090	2 400
19	3/4	1 590	3 500	1 450	3 200	1 300	2 860
22	7/8	2 180	4 800	1 880	4 150	1 725	3 800
26	1	2 585	5 700	2 180	4 800	2 540	5 600
28	1 1/8	3 630	8 000	3 175	7 000	–	–
32	1 1/4	3 990	8 800	3 630	8 000	–	–
38	1 1/2	5 805	12 800	5 625	12 400	–	–
40	1 5/8	7 260	16 000	6 350	14 000	–	–
44	1 3/4	8 800	19 400	8 165	18 000	–	–
52	2	10 705	23 600	9 070	20 000	–	–

See also BS.4928 (Man-made Fibre Ropes)

FIBRE ROPE

Since rope on a job is rarely new, the lifting crew will have to use judgment as to what value to use. If there is any doubt as to the type or condition of the rope it should not be used at all. There can be no substitute for safety.

Table 2.12 provides the approximate SWL of 2-in-1 braided ropes i.e., braided core and braided cover.

CARE AND USE OF FIBRE ROPE

COILING AND UNCOILING

Fibre rope, like wire rope, can be damaged while it is being removed from the shipping coil. If you are handling a new coil of rope, lay it flat on the floor with the inside rope end at the base closest to the floor. Reach down inside the coil and pull the inside rope end up through the coil and unwind it in a counter-clockwise direction (Fig. 2.9). Even when the rope is unwound correctly, loops and kinks may form and these must be carefully removed, otherwise when pulled tight they will cause severe damage to the rope.

After use, the rope should be re-coiled in a clockwise direction, as shown in Fig. 2.10.

Loop the rope over the left arm a number of times until about 5 m (15 ft) of rope remains. When coiling the rope remove kinks as they form. Then, starting about 0.3 of a metre or one foot from the top of the coil, the rope should be wrapped about six times around the loops by rolling them in the left hand.

Then the left hand is extended through the coil and the bight pulled back through the loops. Two half-hitches are tied around the bight, leaving a short end for carrying or for tying to a peg or supporting bar.

WHIPPING

Whenever a fibre rope is cut, the rope ends must be bound or whipped to prevent the rope from untwisting and fraying and the strands from slipping in relation to each other thereby causing one of them to assume more or less than its share of the load. Each of these conditions results in shortened rope life. (Fig. 2.11)

Ordinary whippings are made with fine twine as follows:

Make a loop in the end of the twine and place the loop at the end of the rope, as shown in Fig. 2.12(1). Wind the standing part around the rope covering the loop of the whipping, but leave a small loop uncovered, as shown in (2). Pass the remainder of the standing end up through the small loop and pull the dead end of the twine, thus pulling the standing end and the small loop (through which it is threaded) back towards the end of the rope underneath the whipping. Pull the dead end of the twine until the small loop with the standing end through it reaches a point midway underneath the whipping (3). Trim both ends of the twine close to the loops of the whipping.

Fig. 2.9 Correct method of removing rope from a shipping coil

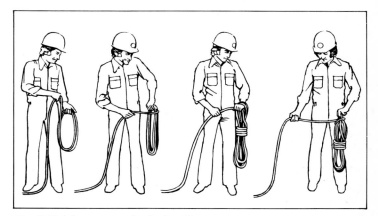

Fig. 2.10 Correct method of coiling a rope

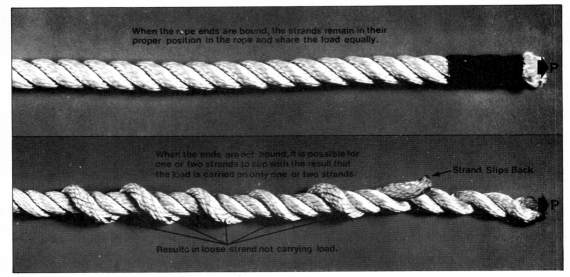

Fig. 2.11 Reason for whipping all rope ends

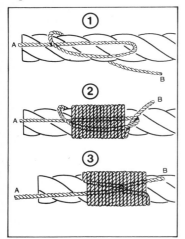

Fig. 2.12 Correct method of applying whipping

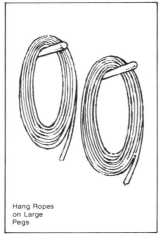

Fig. 2.13 All ropes should be hung up to allow them to dry

Fig. 2.14 All ropes should be cleaned periodically

STORAGE

Poor storage conditions can cause fibre rope to deteriorate as rapidly as harsh usage does. In order to keep ropes in good condition for as long as possible:

— Store them in a dry cool room that has good air circulation. A temperature of 10–20°C (50–70°F) and a humidity of 40–60% are recommended.
— Do not store them on the floor, in boxes, or in cupboards where the air circulation is restricted. They should be hung up in loose coils on large-diameter wooden pegs well above the floor in an area where there is good air circulation. (Fig. 2.13)
— Protect them from the weather, dampness and sunlight. They should be kept away from boilers, radiators, steam pipes and other sources of heat plus all exhaust gases.
— Dry and clean wet ropes before storing them. Moisture not only hastens decay but also causes the rope to kink very easily. If a wet

rope becomes frozen, it must not be disturbed until it is completely thawed, otherwise the frozen fibres will break when handled. Allow wet and frozen rope to dry naturally. Too much heat will cause the fibres to become brittle in a few hours and the rope will be unfit for further service. If a sour odour is detected on natural fibre ropes, the rope should be aired until the odour disappears.

USE

If fibre ropes are abused then you are not only wasting money but also introducing unnecessary hazards to the operation since they will not retain their full capacity.

Remember, the rope that takes your abuse today, may have to take your weight tomorrow. Play it safe. Follow these guidelines to safe rope usage:

- Keep ropes clean and dry. When they become dirty wash them in clean cool water and hang them up to dry. Dirt on the surface or embedded in the rope acts as an abrasive on the strands and fibres. (Fig. 2.14)
- Keep ropes away from chemicals, excessive heat, acids, acid fumes, strong alkali, drying oils, paint, fumes and exhaust gases. They can cause serious damage and strength reductions. Most of the synthetic ropes withstand corrosive chemicals better than natural fibre ropes, but to be certain keep acids, etc. away from *all* ropes. If contamination is suspected wash synthetic ropes in cold water but take natural fibre ropes out of service.
- Never overload a rope. Apply the factor of safety of at least 5:1 or 10:1 and make even further allowances for the age and condition of the rope.
- Never use a frozen rope. It must be carefully handled to avoid breaking the fibres. Allow it to thaw and dry.
- Never drag a rope along the ground. The outside surfaces will become worn and cut by the abrasive action and grit will work into them, become embedded and destroy the internal fibres.
- Never drag a rope over sharp or rough edges and never drag one part of a rope over another part.
- Avoid all but straight line pulls with rope. Abrupt bends interfere with the stress distribution of the fibres that make up the strands. With a straight pull, a rope will give 100% efficiency; tie a knot in the same rope or bend it severely and it is weakened by approximately 50%. (Fig. 2.15)
- Pack all sharp corners when lifting materials with rope. (Fig. 2.16)
- Use thimbles in the eyes of all ropes. When a rope is attached to a hook, ring or pulley block, a thimble should be placed in the loop or eye to reduce the wear on the rope, and to decrease the stresses developed in the rope when it is bent around a very small diameter. (Fig. 2.17)
- Observe proper sling angles. Severe load increases occur as the sling angle is decreased. (Fig. 2.18)
- Never use fibre rope near welding or flame cutting operations. The sparks and molten metal can cut through the rope or set it on fire.
- Avoid unnecessry exposure to strong sunlight. Prolonged exposure will degrade and weaken the rope.
- Avoid exposure to all forms of heat.
- Never couple a left-hand laid rope to a right-hand laid rope.
- When coupling wire and fibre ropes, use metal thimbles in both eyes to prevent the eye of the fibre rope from being cut by the wire rope. (Fig. 2.19)

When blocks are used, the rope must be the correct size for the sheaves. The sheave grooves must be smooth and properly radiused to provide good rope seating. There should be no projections and rough or sharp edges on the blocks which might cut or chafe the rope. The sheaves must have diameters at least six times and preferably ten times greater than the rope diameter.

When using synthetic ropes, extra caution must be taken because:

- Synthetics stretch more than manila.
- Synthetics have a low melting point and are not suitable for use where high temperatures prevail or where friction may cause the rope to melt or fuse.
- Synthetics can be slippery, particularly when new and especially when wet, and as such extra care may be required when tying knots and when handling them.

FIBRE ROPE

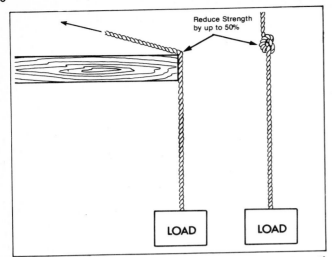

Fig. 2.15 How knots and severe bends affect rope strength

Fig. 2.16 Effective use of corner pads

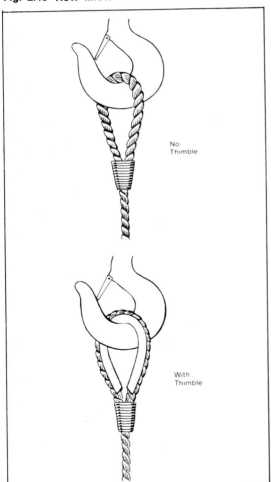

Fig 2.17 Use thimbles in the eyes of all lifting ropes

INSPECTION

The only way to determine the safety of a rope, its life expectancy and its load-carrying ability is by regularly inspecting ever foot of its length. Base your estimate on the section showing the most deterioration.

Care should be taken during this inspection to avoid distorting the lay. The main points to be watched for are external wear and cutting, internal wear between the strands and deterioration of the fibres.

First inspect the outside of the rope. Check for broken fibres and yarns, cuts, nicks, signs of abrasion, burns, unlaying and reductions in diameter. Each represents a loss of strength so their size, number and distribution must be considered when determining if the rope is suitable for its intended use. Acid contact with a manila rope results in dark brown spots, and in synthetic rope, it may either eat away the rope or make the fibres very brittle. (Fig. 2.20)

Open up the rope by untwisting the strands but take care not to kink them. The interior of the rope should be as bright and clean as when it was new. Check for broken yarns, excessively loose strands and yarns, or an accumulation of a powder-like dust, which indicates excessive internal wear between the strands as the rope is flexed back and forth in use. (Fig. 2.21)

If the rope is large enough, open up a strand and try to pull out one of the inside yarns, keeping in mind that if a rope has been overloaded, it is the interior yarns that will have failed first. Excessive oil on the outside of

FIBRE ROPE

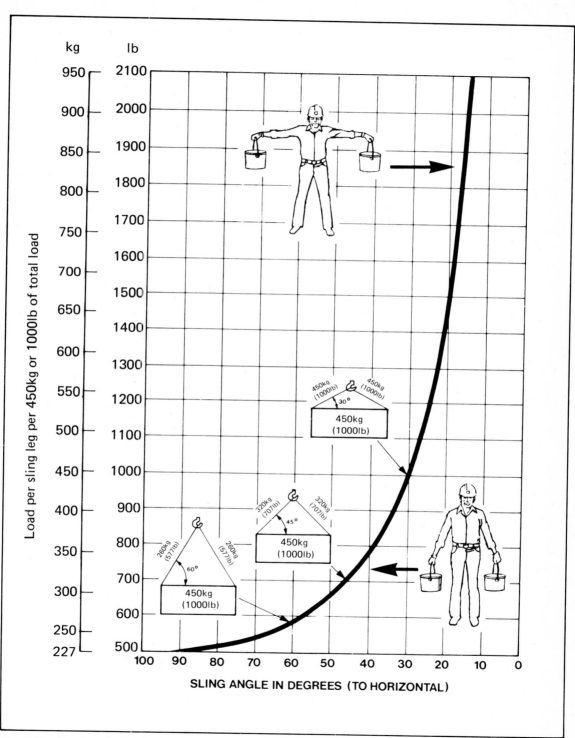

Fig. 2.18 Graph showing how sling angle affects sling load

FIBRE ROPE

Fig. 2.19 Use thimbles in both eyes when connecting fibre and wire ropes

Fig. 2.20 Typical damage to watch for during rope inspection

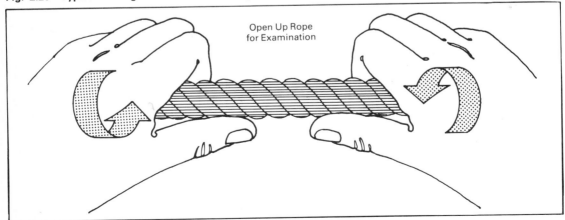

Fig. 2.21 Correct method of opening up a rope

FIBRE ROPE

Fig. 2.22 High-stranded rope

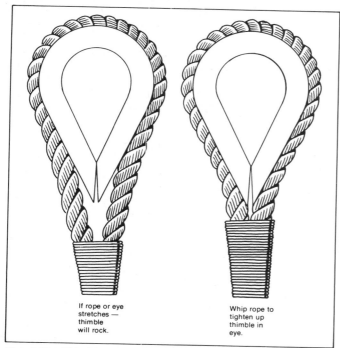

Fig. 2.23 Thimble must fit tightly in eye

Fig. 2.24 All splice tucks must be secure

a new manila rope is also an indication that it has been overloaded.

If it is four-strand rope with a core, try to pull out the heart. If it comes out in short pieces, the rope has been overloaded and should not be used for lifting.

If possible, pull out a couple of long fibres from the end of the rope and try to break them. If they break easily then the rope should be replaced.

If the inside of the rope is dirty, if the strands have begun to unlay, or if the rope has lost its life and elasticity, it should not be used for lifting purposes. If the rope is high-stranded and presents a spiral appearance, or if the heart protrudes, the load will not be equally distributed on the strands and a very short life may be expected. (Fig. 2.22)

Often the surface of a rope feels dry and brittle, or it may show evidence of having been in contact with a hot pipe or other source of heat (glazing, fused sections), or it may be discoloured as the result of exposure to acid fumes. In all cases it should be discarded.

If thimbles are loose in eyes due to rope stretch they can be retightened in the eye by seizing the eye. Never allow a thimble to become so loose that it can rock in the eye. (Fig. 2.23)

Ensure that all splices are properly served or taped. Do not allow any tuck to become undone. Every tuck is necessary for optimum splice efficiency. (Fig. 2.24)

If a rope is weak in just one spot, that portion can be cut out and the rope spliced. A short splice should be used wherever possible, but use a long splice if the rope is to be run through a pulley.

Should there be any doubt as to whether or not a rope is fit for use, it should be replaced at once. Never risk danger to life or damage to property by taking a chance.

When a rope has been condemned, it should be destroyed at once or cut up into short lengths so that it cannot be used for lifting purposes.

SPLICING

A rope splice is the joining of the ends of two ropes or the end of the rope with the body of the rope by weaving the strands over and under the strands of the other part.

Remember that even with a new rope, a splice, regardless of the type, reduces the load-carrying ability of that rope by 10 to 15%.

If synthetic rope is to be spliced, take extra precautions owing to the large number of filaments in each yarn and the smooth surface of the strands. It is also advisable to insert extra tucks to maintain full efficiency and prevent slippage.

Care must be taken during the splicing operation to avoid loss of twist and to maintain the form and lay of yarns and strands. Owing to the smooth cylindrical form of the synthetic fibres, yarns do not 'cling' to each other in the manner of natural-fibre yarns and, accordingly, care must be exercised to prevent the part of the strands used for tucking from separating into what can only be described as a bundle of yarns.

With short and side splices, four full tucks, tapered to finish with one-half an one-quarter tucks will give full efficiency. Where very heavy and rapidly fluctuating loads are involved the number of full tucks can be increased as a safety measure, but with medium-sized and large ropes it is, at all times, advisable to taper the splice in order to obtain maximum strength. Well-made tapered splices on synthetic fibre ropes will yield a strength of approximately 90% of the initial breaking load of the rope.

Here is a general guide to the making of secure splices:

- Care should be taken when unlaying the rope that the formation of the individual strands is not disturbed.
- After the rope has been unlayed, maintain the strand form by applying a series of half-hitches to each strand at intervals of about 25 mm (1 in) with twine.
- Seize the rope at the throat of a side splice to prevent any unlaying during the splicing operation.
- Apply a greater number of tucks in synthetic fibre ropes than you would apply to natural fibre ropes.
- The minimum number of tucks recommended is five. These can be made up of five tucks for small ropes and four full tucks plus two tapered tucks for medium-sized or large ropes.
- The yarn ends of the strands should be left with long tails, preferably not less than 25–50 mm (1–2 in) and simply seized to the finished rope.

When using spliced rope, never subject the splice to a severe bend such as would occur if one were placed in a hook throat. The severe bending and flexing tends to spring the splice.

SPLICES

Short splice: While this is the strongest of all splices, it is not suitable if the rope runs over sheaves or through blocks because the diameter of the rope is almost doubled at the point of joining. They are most suitable for making endless slings, etc.

Fig. 2.25 (1)–(6) is a step-by-step guide to making a short splice, as follows:

(1) Unlay the strands at one end of each rope for six to eight turns. The ends of the strands should be whipped to prevent their untwisting and brought together so that each strand of one rope alternates with a strand of the other rope.
(2) Bring the ends tightly together and apply a temporary seizing where they join.
(3) Take any one strand and begin tucking, the sequence being over one and under one. Strand A is passed over the strand nearest to it, which is D, and then under the next strand, E.
(4) Rotate the splice away from you one-third of a turn and make the second tuck. Strand B is passed over E and then under F.
(5) Before making the third tuck, rotate the splice again one-third of a turn away from

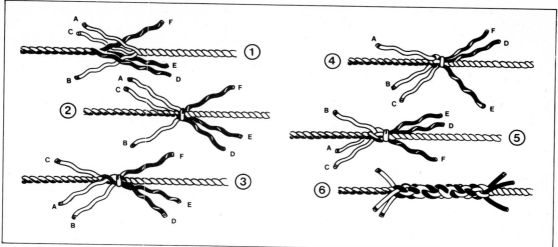

Fig. 2.25 Short splice

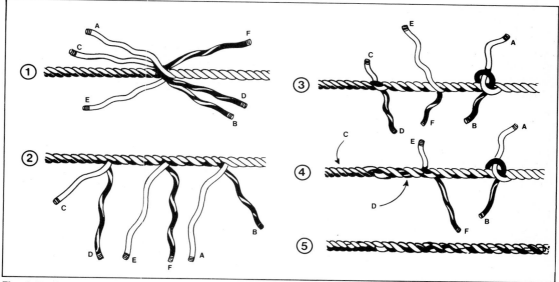

Fig. 2.26 Long splice

you. Strand C is then passed over F and under the next one, D.

This completes the first round of tucks in the left-hand half of the splice. Each strand should now be tucked at least twice more, always over one and under one as before, making sure that each strand lies snug and with no kinks.

To finish the splice, reverse the rope end so that strands D, E and F are now at the left instead of the right and repeat the operation on their side of the rope. Each of the six strands will now have had at least three tucks. A tapered splice is made by taking two more tucks with each strand, cutting away some of the threads from each strand before each extra tuck.

(6) When tucking is finished, remove the centre seizing and cut off the ends of all strands, leaving at least 20 mm (¾ in) on each end. To give a smooth appearance roll the splice back and forth, either under your foot or between two boards.

Long splice: Slightly weaker than the short splice but it increases the rope diameter only slightly and thus allows the rope to run through a sheave or block without obstruction and lessens wear and chafing of the rope fibres at the point of splicing. A long splice should be made only with two ropes of the same size. They are not suitable for endless slings because of their lower strength.

Fig. 2.26 (1)–(5) is a step-by-step procedure for making a long splice as follows:

(1) Begin by unlaying one strand of each rope for 10 or 15 turns. Whip the ends of each strand to prevent untwisting. Then lock the two ropes together by alternating the strands from each end.
(2) Starting at one end, take an opposite pair of strands, A and B, and unlay A. Follow it with B, turn by turn, continuing until only 300 mm (1 ft) or less of B remains. Keep B tight during this step and pull it down firmly into A's former place. Repeat this operation with C and D. Strand D is unlaid and C is laid in its place.
(3) Now each pair of strands is tied loosely together with a simple overhand knot, as indicated by strands A and B. Each knot is then pulled down into the rope like C and D.
(4) Each strand is now tucked twice, over and under, as in the short splice. The illustration shows strands C and D after tucking. If a smaller diameter splice is desired, taper by tucking each strand twice more, cutting away some of the threads for each additional tuck.
(5) When tucking is finished, cut off all strands close to the rope and roll the splice on the floor under your foot or between two boards to smooth it out.

Side (or eye) splice: This splice is normally used to form an eye in the end of a fibre rope. The side splice is also called the eye splice because it is used to form an eye or loop in the end of a rope by splicing the end back into its own side. This splice is made like the short splice except that only one rope is used. Metal or nylon thimbles should be fitted to all side splices for lifting.

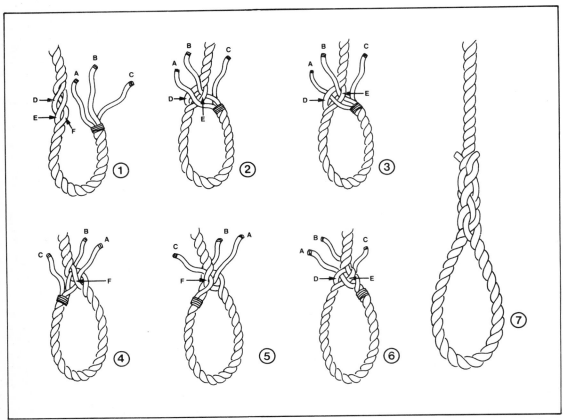

Fig. 2.27 Side (or eye) splice

FIBRE ROPE

Fig. 2.27 (1)–(7) is a step-by-step guide, as follows:

(1) Start by seizing the working end of the rope. Unlay the three strands A, B and C to the seizing and whip the end of each strand. Then twist the rope slightly to open up strands D, E and F of the standing part of the rope as indicated.
(2) The first tuck is as shown in the illustration. The middle strand is always tucked first, so strand B is tucked under E, the middle strand of the standing part.
(3) The second tuck is now made. Left strand A of the working end is tucked under D, passing over E.
(4) This figure shows how the third tuck is made. In order to make strand F easy to get at the rope is turned over. Strand C now appears on the left side.
(5) Strand C is then passed to the right of and tucked under F, as shown. This completes the first round of tucks.
(6) This illustration shows the second round of tucks started, with the rope reversed again for ease in handling. Strand B is passed over D and tucked under the next strand to the left. Continue with A and C, tucking over one strand and then under one to the left. To complete the splice, tuck each strand once more.
(7) The finished side splice is as shown. Remove the temporary seizing and cut off the strand ends, leaving at least 13 mm (½ in) on each end. Roll the splice back and forth under your foot or between two boards to even up and smooth out the strands.

Where metal thimbles are used in the eyes it is important to have a correct design. Synthetic fibre ropes are more extensible under load than natural fibre ropes, and under load the side splice pulls a greater distance from the thimble end. This can result in the angular sides of the thimble losing contact with the rope at the eye. Accordingly the only bearing of rope to thimble is around the thimble radius and as a result rocking of the thimble can occur, leading to chafing of the rope inside the eye, and ultimate breakdown. Seize the rope to prevent this.

Flemish eye splice: While taking longer to make, this splice affords greater resistance to splice failure or slipping than does the side or eye splice.

Fig. 2.28 shows in four steps how this splice is made, as follows:

(1) Unlay one strand of the rope a distance sufficient to account for the eye and splicing.
(2) Form a loop with the two unlaid strands.
(3) Lay the first strand back in its former place in the rope but work around the other side of the eye.
(4) Terminate the splice by making an ordinary side or eye splice as previously described.

Crown knot and back splice: These are used to prevent the strands at the end of a rope from unlaying. Fig. 2.29 shows how these are developed, as follows:

(1) The illustration shows the crowning begun. Open the strands of the rope as shown, lay strand A down over the centre of the rope, bring B down over A and C over B and through the bight of A.
(2) The illustration shows how the strands tie when they are nearly taut. Pull the crown tight then pass one end over the next strand in the standing part and under the following one.
(3) Do the same to each of the other strands in succession, three times, then cut off the ends. This completes the back splice.

KNOTS, BENDS AND HITCHES

The distinction between knots, bends and hitches is generally accepted as being as follows: a *knot* is the intertwining of the end of a rope within a portion of the rope; a *bend* is the intertwining of the ends of two ropes or of the same rope to make one continuous rope or endless rope; a *hitch* is the attachment of a rope to a post, pole, ring, hook or other object.

Hundreds of different knots, bends and hitches have been devised to suit varying purposes, but of this number some have been found to be superior for construction lifting equipment and these are the only ones considered in this manual.

A good knot, bend or hitch is one that can be tied with speed and ease and which, when tied,

FIBRE ROPE

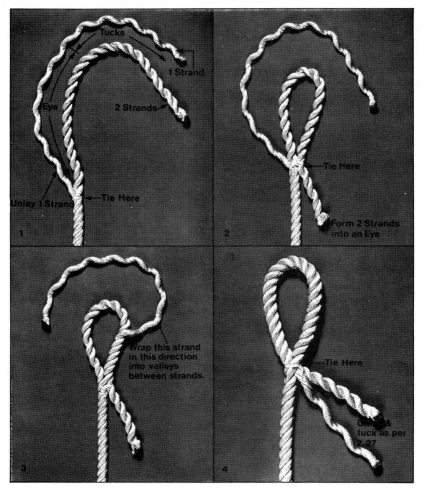

Fig. 2.28 Flemish eye splice

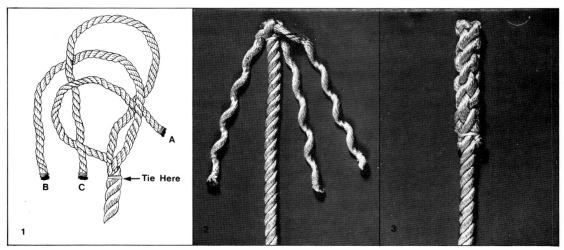

Fig. 2.29 Crown knot and back splice

FIBRE ROPE

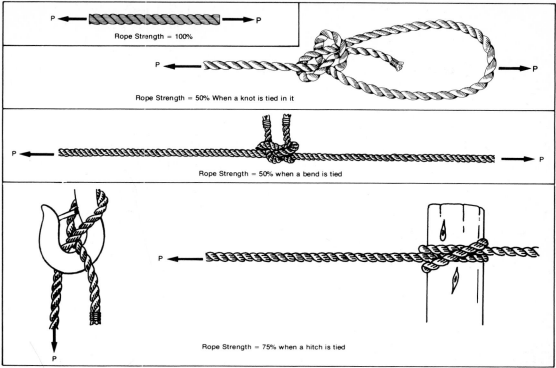

Fig. 2.30 How knots, bends and hitches affect rope strength

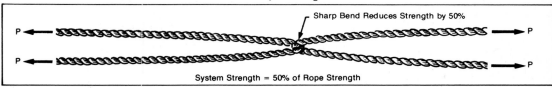

Fig. 2.31 Capacity of two ropes looped together

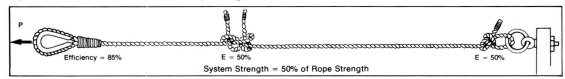

Fig. 2.32 System strength in a combination of knots and splices

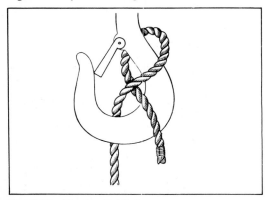

Fig. 2.33 Blackwall hitch

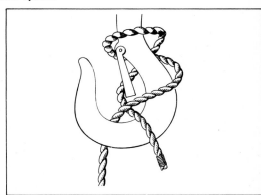

Fig. 2.34 Double blackwall hitch

will hold. Their prime requirements are security against slipping, suitability and strength.

A suitable fastening (knot, bend or hitch) must be selected for the job to be done and it must be tied correctly to obtain the maximum strength and security.

Remember that a rope fastening is never as strong as the original rope because knots, bends and hitches reduce its strength. The bending of the rope in the making of a knot or hitch causes the outside fibres to carry more than their share of the loads and the resultant stretching of the fibres weakens the rope. When a failure occurs, the outside fibres are first to break, followed by the inside fibres. **Knots and bends cut the rope strength by 50% and hitches reduce the strength by 25%.** These figures relate to the ultimate strength of the rope at the knot and must not be accepted as representative of the holding power of the knot against slip in certain applications. (Fig. 2.30)

Loops formed by one rope passing over another of the same size will yield the full strength of the single rope, i.e., the breaking load of the system is no greater than that of a single rope. (Fig. 2.31)

It is important to note that the strength losses due to splices and knots are not cumulative in an assembly. For example, if a load is being carried by two ropes of equal size, joined by a Carrick bend (Fig. 2.40) and anchored to one end with a side splice and the other with a bowline (Fig. 2.35), the strength of the assembly is, in fact, the strength of the bowline, i.e. 50% of the tensile strength of the ropes. (Fig. 2.32)

Unless absolutely necessary, do not use knotted ropes for overhead lifting, and when using synthetic fibre ropes it is advisable to double knot because of the rope's smooth exterior.

All the common knots can be tied successfully in synthetic fibre ropes, but in regard to polyethylene and, to a certain extent, polypropylene monofilament ropes, most knots tend to slip and it is necessary to 'double up' on the knot in order to hold. This is due to the waxy nature of the monofilament surfaces.

Blackwall hitch: (Fig. 2.33) Satisfactory for temporarily attaching light loads to a hook and only if the tension is never allowed to slacken and the rope is dry. It consists of a loop passed under the standing part and across the hook. Under load, the live end jams and holds the dead end against the hook. Synthetic ropes are too slippery for this hitch.

Double Blackwall hitch: (Fig. 2.34) Provides better security than the single Blackwall and can be used with wet or slippery ropes.

Bowline: (Fig. 2.35) Never jams or slips if correctly tied. It is a universal knot that is easily tied and untied. Two interlocking bowlines can be used to join two ropes together, single bowlines can be used for lifting or hitching directly round a ring or post. Make an overhand loop with the end of the rope held towards you. Pass the end of the rope up through the loop, then up behind the standing part and back down through the loop. Draw up tight.

Double bowline: (Fig. 2.36) Forms a good non-slipping eye or set of spreaders for a sling. Make an overhand loop with the end of the rope held towards you, exactly as for the ordinary bowline. Pass the end through the loop twice – making two lower loops. The end is then passed behind the standing part and down through the first loop again as in the ordinary bowline. Pull tight.

Bowline on the bight: (Fig. 2.37) The knot is used to tie a bowline (Fig. 2.35) in the middle of a line or for making up a set of double leg spreaders. It can also be used as a sling to sit in by sitting in one loop and pulling the other loop around the back and under the arms. To tie, double the rope, make an overhand loop and pass the loop end up through it. Bring the loop end towards you, down and over the other loop section. Bring the loop up back until it lies behind the standing part. Draw it tight by a slow even pull on the upper right side as shown.

Triple (rescue) bowline: (Fig. 2.38) Is effective for lifting injured persons since a third loop is available to support the back and shoulders as well as individual loops for each leg. Double the rope, make an overhand loop and pass the loop end up through it then up and in behind the standing part and back down through the loop, thus making the third loop. Draw up tight.

Running bowline: (Fig. 2.39) Makes an excellent slip knot or noose. It runs freely on the standing part, is easily untied and is very

FIBRE ROPE

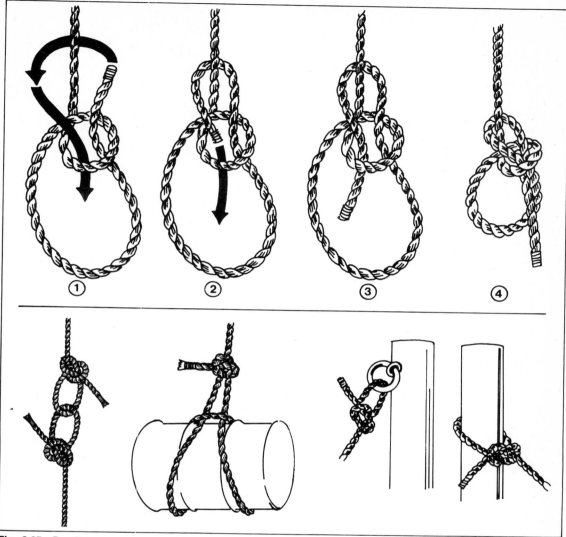

Fig. 2.35 Bowline

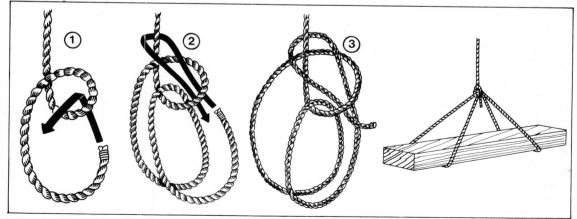

Fig. 2.36 Double bowline

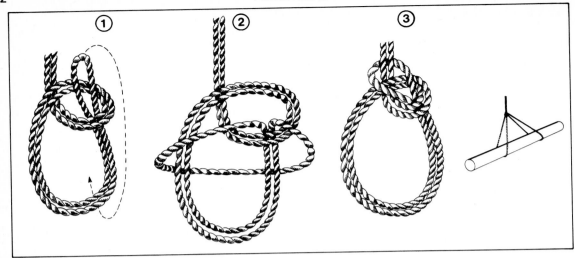
Fig. 2.37 Bowline on a bight

Fig. 2.38 Triple (rescue) bowline

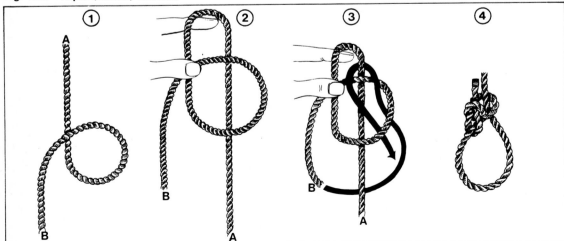
Fig. 2.39 Running bowline

FIBRE ROPE

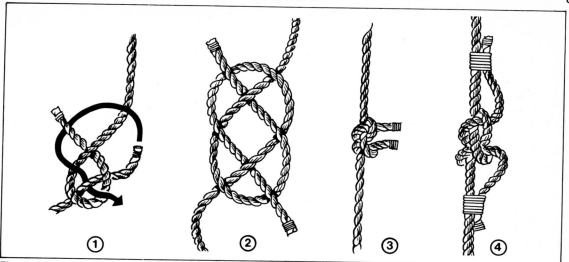

Fig. 2.40 Carrick bend

Fig. 2.41 Catspaw

Fig. 2.42 Clove hitch

Fig. 2.43 Figure eight knot

strong and efficient. To tie, make an overhand loop with the end of the rope held towards you, hold the loop with your thumb and fingers and bring the standing part of the rope back so that it lies behind the loop. Take the end of the rope in behind the standing part, bring it up and feed it through the loop. Pass it behind the standing part at the top of the loop and bring it back down through the loop.

Carrick bend: (Fig. 2.40) This knot can be used for joining ropes together and is easier to untie than most knots after being subjected to load. It is one of the strongest knots, cannot jam and under load it always draws up tight. This is important because very heavy ropes usually cannot be fully tightened by hand. With one rope end form an underhand loop with both the free end and standing part pointing away from you. Start the second rope end beneath both sides of the loop. Cross it over the standing part of the first rope. Then under the free end of the first rope. Then over the left side of the loop. Cross it under itself and let the second free end lie over the right side of the loop. Finish by seizing each end to the standing part.

Catspaw: (Fig. 2.41) Is used to attach a rope to a hook and is particularly useful when securing the middle of a rope to a hook. It does not jam and unties by itself when removed from the hook. It is tied by forming two loops in the rope and twisting them away from you (at least three twists). The two twisted loops are then brought together and placed over the hook.

Clove hitch: (Fig. 2.42) The clove hitch is a quick, simple method of fastening a rope around a post or pipe. It can be tied in the middle or at the end of a rope. But since it has a tendency to slip when used at the end of a rope, the end should be half-hitched (Fig. 2.44) to the standing part for greater security. The hitch is tied by forming an underhand loop around the post followed by another underhand loop, as shown.

Figure eight knot: (Fig. 2.43) Generally tied at the end of a rope as a temporary measure to prevent the strands from unlaying. It can be tied simply and quickly, and does not jam as easily as the overhand knot. It is also larger, stronger and does not injure the rope fibres. It is useful in preventing the end of a rope from slipping through a block or an eye. To tie, make an underhand loop, bring the end around and over the standing part, and pass the end under and then through the loop. Draw up tight.

Half-hitch: (Fig. 2.44) The half-hitch is generally used for fastening a rope to an object for a right-angle pull. To tie, pass the end of the rope around the object and tie an overhand knot to the standing part. The illustration shows the half-hitch tied with the end nipped under the turn of the rope some distance away from the

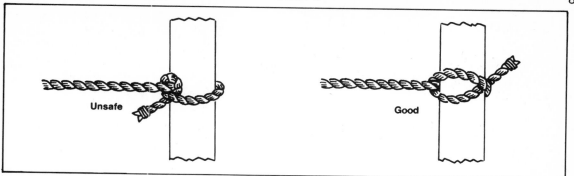

Fig. 2.44 Half-hitch

Fig. 2.45 Two-half-hitches

standing part; this method is fairly reliable for temporary use if the pull is steady and the arrangement is not disturbed.

Two-half-hitches: (Fig. 2.45) Just as its name states, two-half-hitches is simply a half-hitch tied twice. It is quickly tied, reliable, and can be put to almost any general use.

Noose (or halter) hitch: (Fig. 2.46) Can be used to tighten a rope around an object. Tie as shown in diagram, making sure that final tuck is made with bight near end. Then after drawing up, drop the end through this bight so that knot cannot fail.

Overhead knot: (Fig. 2.47) Can be used for joining two ropes. It will not slip but tends to jam and has a low efficiency.

Reef (or square) knot: (Fig. 2.48) Can be used for tying two ropes of the same diameter

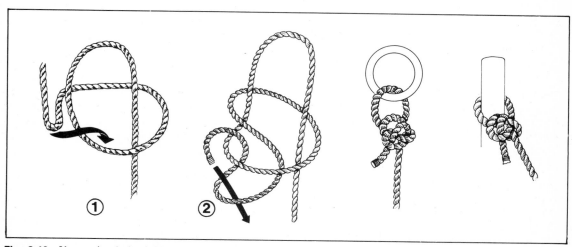

Fig. 2.46 Noose (or halter) hitch

Fig. 2.47 Overhead knot

together. It is unsuitable for wet or slippery ropes and should be used with caution since it unties easily when either free end is jerked. The knot is easy to tie but the finished knot should be checked to make certain it is correct; both the live and dead ends of the ropes must come out of the loops on the same side.

Rolling hitch: (Fig. 2.49) This knot can be effectively used for lifting round objects, such as pipe or bar steel, and for fastening a rope to another rope or a pole parallel to it. It grips tightly if the pull is constant but slackens when the load is released.

Round turn and two-half-hitches: (Fig. 2.50) Can be used to secure a rope to a column or post. It is easily tied, does not jam, and will stand heavy strains without slipping.

Sheet bend: (Fig. 2.51) Can be used for joining ropes of different sizes provided they are of light or medium size. On large ropes the Carrick bend (Fig. 2.40) is preferable. To tie, make an overhand loop with the end of one rope. Pass the end of the other rope through the loop thus formed, then up behind its standing part, then down through the loop again. Draw up tight.

Sheepshank: (Fig. 2.52) The sheepshank is intended to shorten a rope for temporary use only. Carefully tied and drawn up tight, it is fairly reliable under a steady pull. It is especially useful if the ends of the rope are not free as it can be tied anywhere in the length of an already-tied rope. It can also be used to take the strain off a damaged piece of rope when there is not time to replace it. The knot can be made more secure if it is seized. To tie, form an S loop, as shown in the diagram, then with one free end of the rope make a half-hitch (Fig 2.44) and slip it over one of the loops. Tighten. Repeat procedure with the other loop.

Slippery hitch: (Fig. 2.53) Provides a quickly-tied and easily released hitch for securing a rope to rings, posts or light loads.

Snubber: (Fig. 2.54) For holding or slowly lowering a heavy load. The load on the hand line is only a fraction of the load on the live line. Three or four turns should be taken.

Timber hitch: (Fig. 2.55) Generally used for fastening rope to posts and for hoisting planks, timbers, pipe, etc. but use it only when the load is steady. The timber hitch holds without slipping and does not jam but loosens when strain is relieved. One or two half-hitches (Fig. 2.44) should be added to keep a plank or length of pipe on end while it is being lifted. To tie, pass a rope around the object and take a turn with the end around the standing part. Then, as shown in the diagram, twist or turn the end back on itself. Three turns back are sufficient and they should follow the lay of the rope.

Triple sliding hitch: (Fig. 2.56) Is generally used to secure a safety belt lanyard to a life line. Manila rope must not be used for the lanyard material because if the person wearing it were to fall, the knot would grab the line and create shock loads and severe bending of the rope such as would cause manila rope to fail. Use nylon for all lanyards.

Scaffold hitch: (Fig. 2.57) The scaffold hitch is used for fastening single scaffold planks and needle beams so that they hang level and are prevented from tilting.

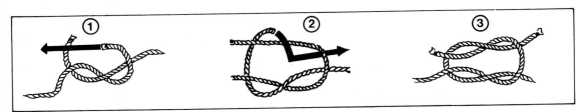

Fig. 2.48 Reef (or square) knot

FIBRE ROPE

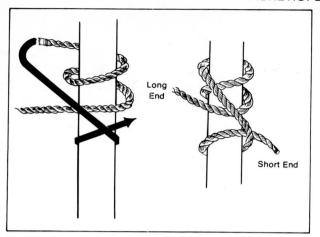

Fig. 2.49 Rolling hitch

Fig. 2.50 Round turn and two-half-hitches

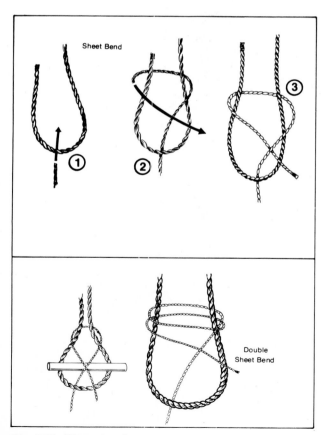

Fig. 2.51 Sheet bend

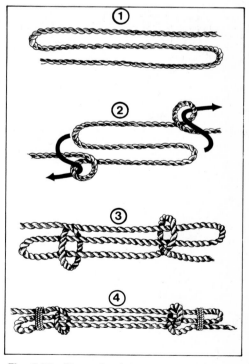

Fig. 2.52 Sheepshank

FIBRE ROPE

Fig. 2.53 Slippery hitch

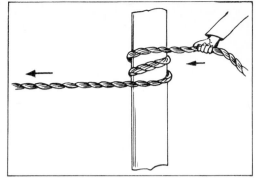

Fig. 2.54 Snubber

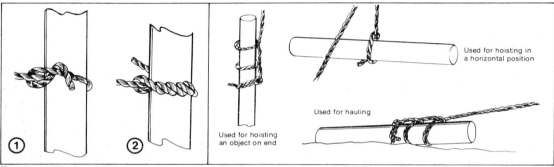

Fig. 2.55 Timber hitch

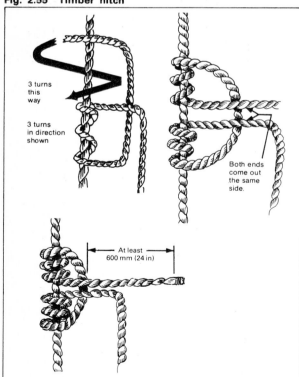

Fig. 2.56 Triple sliding hitch

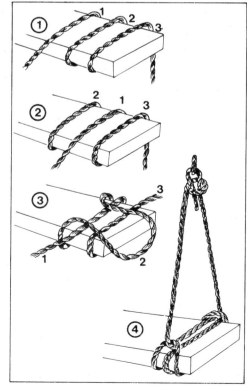

Fig. 2.57 Scaffold hitch

CHAPTER 3

Chain

Chain lifting tackle is an alternative to wire rope tackle. Each has advantages and limitations and the latter must be properly appreciated if failures and accidents are to be avoided.

Compared with wire rope, chain tackle will better withstand rough handling, is less liable to tangle and knotting and is flexible when dead. Chain slings can grip a load firmly and are not damaged as easily as a wire rope sling by sharp corners on a load. Chains are resistant to abrasion and corrosion and will give warning of excessive loading by the elongation and narrowing of the links until sometimes they bind on each other. Wear on the links is easily detected and can be measured to assess loss of strength.

Chain, however, can fail suddenly if one link has an undetected faulty weld, and unless properly marked it is difficult to assess the safe working load (SWL) and the quality of the steel.

When a wire rope is fatigued from severe service, the wires break one after the other over a relatively long period of time and thus afford the inspector an opportunity to discover the condition. If severely overloaded, the wires and strands will break progressively over a period of perhaps several seconds and with considerable noise, before complete failure occurs. This may afford the man handling the load a brief time in which to either lower it or jump to safety before it falls.

GRADES

All chain suitable for construction use and lifting must be marked with the SWL and identity symbols. BS.3458, *Alloy Steel Chain Slings,* requires these marks to be permanently and legibly stamped on the upper terminal fitting, on a solid link, or an idle link; the marks must not be on or near a weld. The symbols must provide identification with the maker's test certificate. In Canada the chain used should be alloy steel which is marked with a letter A on every link. (Fig. 3.1)

Alloy steel chains must never be welded or exposed to excessive temperatures as the results of heat treatment will be affected and the chain will lose its strength.

TABLE 3.1

MAXIMUM SAFE WORKING LOAD ALLOY STEEL CHAIN – BS.3458 Single vertical leg		
Chain link size		Safe load (half specified proof load) tonnes
mm	in	
6	¼	0.76
10	⅜	1.68
13	½	3.05
16	⅝	4.72
19	¾	6.86
22	⅞	9.30
25	1	12.19
29	1⅛	15.39
32	1¼	19.05
35	1⅜	23.01
38	1½	27.45

STRENGTH

Table 3.1 gives the maximum SWL of new alloy steel chain when used as a single vertical sling. Maximum SWL of inclined slings, etc., are presented in the section on Slings.

Under normal conditions the SWL should not exceed one-half the proof load, but when conditions are hazardous or severe the SWL should be substantially less. Where, for example, there is special risk of injury a SWL of not more than $9\,d^2$ should be adopted, or where the links are subjected to severe local pressure, such as bearing on a hard irregular surface, the SWL should not exceed $6\,d^2$.

Any change in the above factors, such as twisting of the chain, deterioration of the chain or its fittings by overload, use, rust, shock loading or inclined (offset) loading will lessen the load that the chain will safely carry.

Alloy steel chain can be used in environments of up to 275°C (500°F) without reducing the SWL but at temperatures above this level the load limit decreases as per Table 3.2.

INSPECTION AND EXAMINATION

All chains used regularly should be thoroughly inspected and examined link-by-link at least once a month by a competent person and in no circumstances should a chain be used for lifting purposes unless it has been closely examined for defects or wear.

Whenever a chain is subjected to shock or impact loads it must be immediately inspected before being put back into service.

It is recommended that every chain carries a small metal identification tag bearing a serial number and its SWL (Fig. 3.2). A log book should also be kept for the chain identifying its characteristics and setting up an inspection and examination schedule.

TABLE 3.2

EFFECT OF HEAT ON SAFE WORKING LOAD (SWL)

Chain temperature °C	°F	Reduction of SWL limit while heated*	Permanent reduction in SWL limit**
275	500	None	None
330	600	10%	None
390	700	20%	None
445	800	30%	None
500	900	40%	10%
550	1000	50%	15%

* While chain is at temperature in first column.
** When chain is used at room temperature after being used at temperature shown in first column.

TABLE 3.3

CORRECTION TABLE TO COMPENSATE FOR WEAR

Nominal chain link diameter		Reduce rated capacity by following % when diameter at worn section is as follows				Remove from service when diameter is	
		5%		10%			
mm	in	mm	in	mm	in	mm	in
6	¼ = 0.250	5.75	0.244	5.70	0.237	5.40	0.223
9	⅜ = 0.375	8.78	0.366	8.55	0.356	8.10	0.335
13	½ = 0.500	12.68	0.487	12.35	0.474	11.70	0.448
16	⅝ = 0.625	15.60	0.609	15.20	0.593	14.40	0.559
19	¾ = 0.750	18.53	0.731	18.05	0.711	17.10	0.671
22	⅞ = 0.875	21.45	0.853	21.45	0.830	19.80	0.783
25	1 = 1.000	24.38	0.975	24.38	0.949	22.50	0.895
29	1⅛ = 1.125	28.28	1.100	27.55	1.070	26.10	1.010
32	1¼ = 1.250	31.20	1.220	30.40	1.190	28.80	1.120
35	1⅜ = 1.375	34.13	1.340	33.25	1.310	31.50	1.230
38	1½ = 1.500	37.05	1.460	36.10	1.430	34.20	1.340
41	1⅝ = 1.625	39.98	1.590	38.95	1.540	36.90	1.450
44	1¾ = 1.750	42.90	1.710	41.80	1.660	39.60	1.570
48	1⅞ = 1.875	46.80	1.830	45.60	1.780	43.20	1.680
51	2 = 2.000	49.75	1.950	48.45	1.900	45.90	1.790

CHAIN

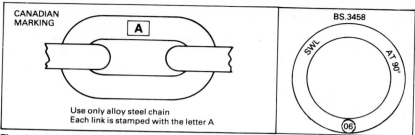

Fig. 3.1 Identification of alloy steel chain

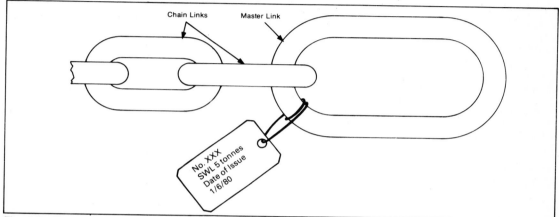

Fig. 3.2 Every chain should have an identification tag

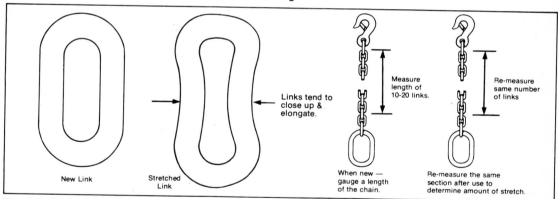

Fig. 3.3 During examinations, look for chain stretch

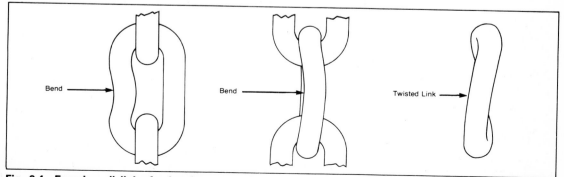

Fig. 3.4 Examine all links for bends, twists or other damage

When inspecting chain remember to give every link a close examination, one bad link is all it needs to fail. The following guidelines may be of assistance in carrying out this work:

— Clean the chain thoroughly in a solvent solution.
— Lay it out on a clean surface or hang it up in a well lit area. Use a magnifying glass to aid in the examination.
— Look for elongated or stretched links. When the links are severely stretched they tend to close up so that in some cases stretch may be indicated by links that bind or a chain that will not hang perfectly straight. Elongation should be determined by measuring all new chains in sections of 300–900 mm (12–36 in) with a calliper and re-measuring them during the inspection. If the inspection reveals a stretch of more than 3%, take the chain out of service. (Fig. 3.3)
— Look for bent, twisted or damaged links that often occur when the sling is used to lift a load which has unprotected sharp edges. (Fig. 3.4)
— Look for cracked links. The presence of any crack, regardless of its size, means the chain is unsafe and must be removed from service. Where a crack is suspected, the link may be soaked in thin oil and then wiped dry. A coating of powdered chalk or other white material is applied to the surface and allowed to remain there for several hours. If a crack exists, the oil pocketed in it will be drawn out and discolour the white coating.
— Look for gouges, chips, scores or cuts in each link. If they are deep or large in area then the chain should be removed from service. If the depth of these defects is such that the link's size is reduced below that listed in the Table for wear (3.3), then the chain is unsafe. In addition, watch all sharp nicks and cuts because cracks usually initiate at or near them. (Fig. 3.5)
— Look for small dents, peen marks and bright polished surfaces on the links. These usually indicate that the chain has been work-hardened or fatigued.
— Look for lifted fins at welds. They are evidence of severe overloading and indicate

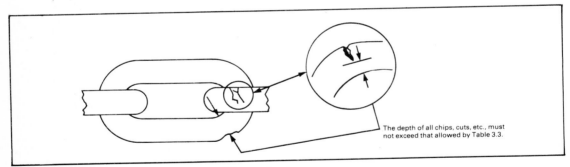

Fig. 3.5 Examine all links for gouges, chips and cuts

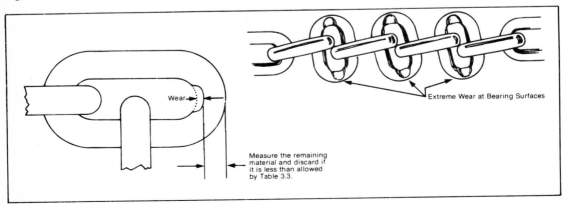

Fig. 3.6 Examine all links for wear at bearing surfaces

CHAIN

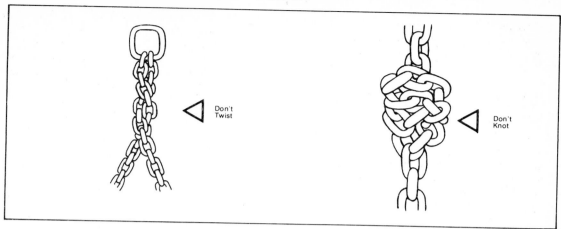

Fig. 3.7 Never twist or knot a chain

that the whole chain should be destroyed.
- Look for severe corrosion resulting in measurable material loss or severe pitting.
- Be particularly careful in determining link wear at the point where they bear on each other. A calliper should be used for the measurement, and the degree of wear at the worst section or the most badly worn link must be determined during this work. Table 3.3 gives the percentages by which the rated loads must be reduced to compensate for wear and the point at which the chain must be removed from service. (Fig. 3.6)

CARE AND USE

- Use only alloy steel chains and never exceed their rated SWL.
- Inspect them regularly and take them out of service and destroy them when they are defective. Damaged or worn sections can sometimes be returned to the manufacturer for repair and reconditioning. Only specialists can repair alloy steel chain — don't attempt it yourself.
- Know the weight of all loads to avoid accidental overload.
- Avoid impact loading.
- Store chains where they will not be damaged or corroded.
- Never shorten a chain by twisting or knotting it or with nuts and bolts. (Fig. 3.7)
- When wrapping chain around sharp corners use pads to prevent damage to the links.
- Never use home-made links or makeshift fasteners formed from bolts, rods, etc.
- Never use repair links, mechanical coupling links or low-carbon steel repair links to splice broken lengths of alloy steel chain. They are much weaker than the normal chain links.
- Never use a chain when the links are locked, stretched or without free movement. Stretching can be distinguished by small checks or cracks in the links, elongation of the links or a tendency for the links to bind on each other.
- Never hammer a chain either to straighten the links or to force the links into position.
- Avoid crossing, twisting, kinking or knotting a chain.
- Never use the tip of a chain hook to carry a load.
- Never re-weld alloy steel chain links. They must be replaced by the manufacturer.
- Inspect each link regularly for wear, nicks, gouges, stretch, localised bending and shearing.
- Make sure the chain is of the correct size and grade for the load.
- Make sure all attachments and fittings are of a type, grade and size suitable for service with the chain used.
- Make sure that alloy steel chains are never annealed or heat treated. Their capacity will be destroyed if they are.

CHAPTER 4

Other equipment

The equipment used with lifting tackle and in its applications is just as important as the ropes, chains, etc., that are used with it. Not only must you know what equipment to use and how to use it, but you must know how the safe working loads (SWL) compare with the rope or chain, etc.

It is extremely important that all fittings be of adequate strength for the application and it is strongly recommended that only forged 'load rated' ('Safe Working Load' stamped on the fitting) equipment be used for overhead lifting.

DRUMS

With regard to drum assemblies, be certain that:

– They have adequate power and operational characteristics to perform all lifting, holding and lowering functions when operated under all conditions and configurations as recommended and approved by the manufacturer.
– They are provided with suitable clutching or power engaging devices that facilitate immediate starting and stopping of the drum motion.
– They are provided with self-setting brakes that are capable of supporting all rated loads with recommended reeving.
– Their brakes and clutches are provided with adjustments to compensate for wear and maintain adequate force in springs where used.
– The drums have sufficient rope capacity with recommended rope size and reeving to perform all lifting and lowering functions under recommended and actual service conditions. In addition, all hoist drums should be provided with adequate means to ensure even spooling of the rope on the drum.
– At least two or three full wraps of rope remain on the drum in all service conditions.
– The drum end of the rope is anchored by a clamp, securely attached to the drums with an arrangement approved by the manufacturer.
– The drums are provided with rims and flange guards of size sufficient to prevent the rope from jumping off the drum.
– Grooved drums have the correct groove pitch for the diameter of the rope. The depth of the groove must also be correct for the diameter of the rope. (Fig. 4.1)
– The flanges on grooved drums must project either twice the rope diameter, or 50 mm (2 in), beyond the last layer of rope, whichever of the two is the greater. (Fig. 4.2)
– The flanges on smooth drums must project either twice the rope diameter, or 63 mm (2½ in), beyond the last layer of rope, whichever of the two is the greater. (Fig. 4.2)
– The fleet angle for smooth drums should lie between ¼° and 1¼° and for grooved drums it should lie between 1° and 2°. (Figs. 4.3, 4.4). BS.302 advises 1½° and 2½°, BS.1757 1 in 12 (4.75°) for all drums.
– Where the operator cannot see the drum or rope, drum rotation indicators should be provided and located to afford easy checking by the operator.

OTHER EQUIPMENT

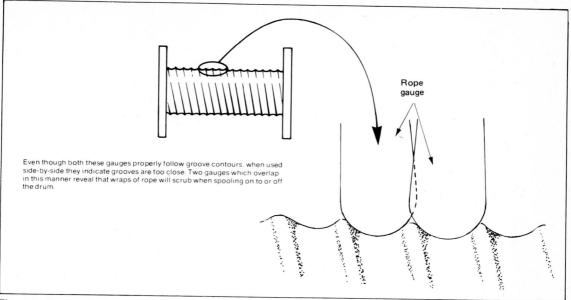

Fig. 4.1 During inspections, check drum grooves

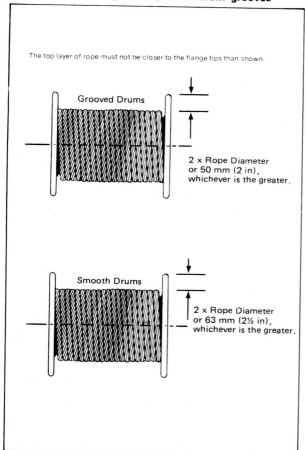

Fig. 4.2 Maximum drum capacity

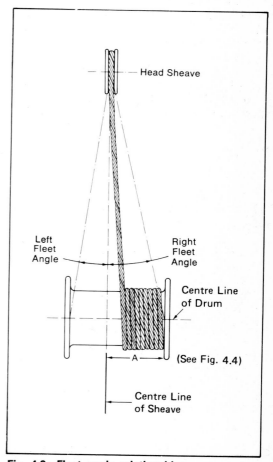

Fig. 4.3 Fleet angle relationships

OTHER EQUIPMENT

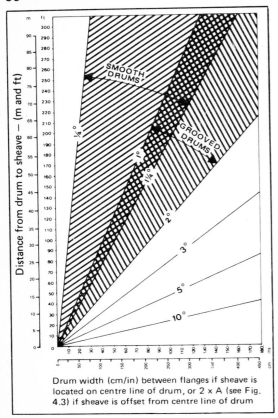

Fig. 4.4 Correct fleet angle ranges for grooved and smooth drums

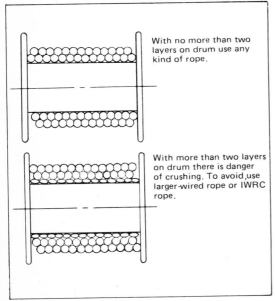

Fig. 4.5 Crushed, jammed and flattened strands

Whenever possible, a drum should be designed to hold all the rope on one smooth, even layer. Two and sometimes three layers are permitted, but more than three layers may cause crushing of the rope on the bottom layer as well as at the end of any layer where pinching occurs. (Fig. 4.5)

Make sure the rope is correctly installed on the drum. The method to be followed is described in Chapter 1.

The manner in which rope spools on a drum depends upon:

- The rope size and construction.
- The drum size and ratio of drum diameter to rope diameter.
- The rope speed.
- The amount of rope to be spooled.
- The type of lifting tackle.
- The load on the rope.

To be certain that the rope spools evenly on the drum, either use a spooling device or keep the fleet angle within the correct limits, and keep tension on the rope at all times. Approximately 10% of the working load is recommended as minimum.

It is often necessary to know the approximate capacity of a given drum or reel for a particular rope diameter. Using Imperial units the following method (Fig. 4.6) can be used to make this calculation:

Add the diameter of the drum (B) to the depth of the flange (A). Multiply this sum by the depth of flange (A). Multiply the result by the distance between the drum flanges (C). Then multiply this result by the factor (F) listed in Table 4.1 opposite the diameter of rope to be installed.

Drum Capacity (Feet of rope) =
$(B + A) \times A \times C \times F$
where A, B and C are in inches.

For example, if the diameter of the drum is 18 in, the depth of the flange is 2 in and the distance between drum flanges is 24 in, the drum's capacity for a ¾ in rope is:

Capacity = $(18 + 2)2 \times 24 \times .465 = 446$ ft.

OTHER EQUIPMENT

TABLE 4.1

DRUM OR REEL CAPACITY FACTORS		
Nominal rope diameter		F*
mm	in	
6	¼	4.160
8	⁵⁄₁₆	2.670
10	⅜	1.860
11	⁷⁄₁₆	1.370
13	½	1.050
14	⁹⁄₁₆	0.828
16	⅝	0.672
19	¾	0.465
22	⅞	0.342
25	1	0.262
29	1⅛	0.207
32	1¼	0.167
35	1⅜	0.138
38	1½	0.116
41	1⅝	0.099
44	1¾	0.085
48	1⅞	0.074
51	2	0.066
54	2⅛	0.058
57	2¼	0.052
60	2⅜	0.046
64	2½	0.042

The factor (F) applies to nominal rope size and level winding. Since new ropes are usually oversized by about 3% of rope diameter, the result obtained by the formula must be decreased to account for oversized rope and/or random or uneven winding, as follows:

For oversized ropes decrease the calculated length by from 0–6%.

For random wound ropes decrease the calculated length by from 0–8%.

Whenever possible use grooved drums rather than smooth drums as the grooves furnish better support for the rope than do flat surfaces, and more uniform winding results in less abrasive wear on the rope. The groove surfaces on grooved drums and the complete surface on smooth drums should be perfectly smooth. Those which have taken the imprint of the outer wires of previous ropes will exert a grinding action on new ropes. (Fig. 4.7)

This imprinting and scoring is caused by high contact pressures between the rope and drum surface. If this condition is evident the drum must be resurfaced and the contact pressure reduced by:

(a) decreasing the load on the rope, or
(b) increasing the drum diameter, or
(c) replacing the drum with one made of harder metal.

The radial contact pressure can be calculated in Imperial units as follows:

$$P = \frac{2L}{Dd}$$

where P = radial pressure in psi (Fig. 4.8)

L = rope load in lb

D = tread diameter of drum or sheaves (inches)

d = rope diameter (inches)

Example: ⅞ in 8 × 19 round strand rope

max. working load = 11 800 lb
drum diameter = 18 in

$$P = \frac{2L}{Dd} = \frac{2(11\,800)}{18 \times ⅞} = 1498 \text{ psi}$$

This contact pressure means, from the Table, that the drum must be manganese steel.

It is suggested that the limits given in Table 4.2 should be observed for all drums and sheaves.

In order to minimise the rope bending stresses the drum diameter should be at least as large as indicated in Table 4.3.

SHEAVES

The condition and contour of sheave grooves have a major influence on rope life. The grooves must be smooth and slightly larger than the rope to prevent it from being pinched or jammed in the groove. Since most ropes are made slightly larger than their nominal size, the sheave grooves for new rope should just accommodate the full oversize of the rope, as given in Table 4.4.

OTHER EQUIPMENT

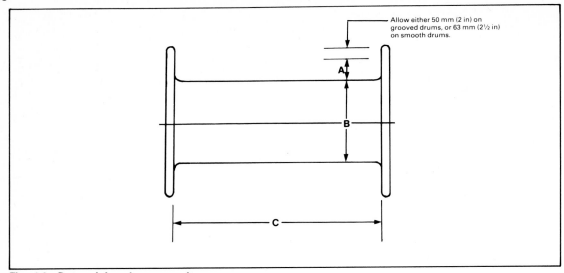

Fig. 4.6 Determining drum capacity

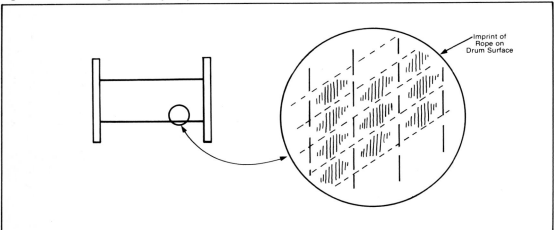

Fig. 4.7 During inspections, check for drum scoring

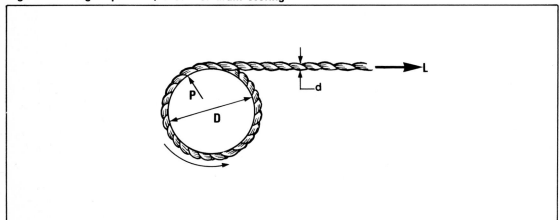

Fig. 4.8 Determination of drum contact pressures

OTHER EQUIPMENT

TABLE 4.2

Rope construction	Cast iron		Cast steel		Manganese steel	
	kg/cm²	lbf/in²	kg/cm²	lbf/in²	kg/cm²	lbf/in²
6 × 7 Reg. Lay	21.0	300	38.5	550	105	1 500
6 × 7 Langs Lay	24.5	350	43.8	625	119	1 700
6 × 19 Reg. Lay	35.0	500	63.0	900	175	2 500
6 × 19 Langs Lay	40.3	575	71.8	1 025	119	2 850
6 × 37 Reg. Lay	42.0	600	75.3	1 075	210	3 000
6 × 37 Langs Lay	49.0	700	87.5	1 250	245	3 500
8 × 19 Reg. Lay		600	75.3	1 075	210	3 000
6 × 8 Flat Strand		500	63.0	900	175	2 500
6 × 25 Flat Strand		800	101.5	1 450	280	4 000
6 × 33 Flat Strand		975	126	1 800	343	4 900

TABLE 4.3

DRUM DIAMETERS (mm/inches) FOR VARIOUS TYPES OF ROPE

Rope diameter		6 × 7		6 × 19 18 × 7 NR		6 × 19 Warr. 6 × 16 F 6 × 27 FS		6 × 19 F 8 × 19 Seale		6 × 37	
mm	in	mm	in	mm	in	mm	in	mm	in	mm	in
6	¼	254	10	203	8	178	7	152	6	127	5
8	⁵⁄₁₆	330	13	279	11	230	9	178	8	152	6
9	⅜	406	16	330	13	279	11	254	10	178	7
11	⁷⁄₁₆	457	18	380	15	330	13	279	11	203	8
13	½	533	21	430	17	380	15	330	13	230	9
14	⁹⁄₁₆	584	23	480	19	430	17	380	15	254	10
16	⅝	635	25	530	21	480	19	406	16	279	11
19	¾	787	31	635	25	560	22	480	19	330	13
22	⅞	939	37	760	30	660	26	584	23	406	16
26	1	1067	42	864	34	760	30	660	26	457	18
28	1⅛	1194	47	965	38	864	34	736	29	508	20
32	1¼	1320	52	1067	42	940	37	813	32	560	22
35	1⅜	1473	58	1194	47	1040	41	914	36	635	25
38	1½	1600	63	1295	51	1140	45	990	39	686	27
40	1⅝	–	–	1400	55	1245	49	1067	42	736	29
44	1¾	–	–	1500	59	1320	52	1140	45	787	31
48	1⅞	–	–	1625	64	1420	56	1245	49	864	34
52	2	–	–	1730	68	1525	60	1320	52	914	36
56	2¼	–	–	–	–	1702	67	1473	58	1016	40
64	2½	–	–	–	–	1905	75	1650	65	1140	45

The bottom of the groove should have an arc of support of at least 120° to 150°, and the sides of the groove should be tangent to the arc. (Fig. 4.9)

The more closely the contour of the groove approaches that of the wire rope the greater becomes the area of contact between the two. This minimises rope distortion, bending fatigue and eases sheave rotation. If the groove diameter is too large, the rope will not be properly supported and will tend to flatten and become distorted. This accelerates bending fatigue in individual wires and can cause premature failures. (Figs. 4.10, 4.11)

If the sheave groove is too narrow for the rope the operating tension will draw the rope deeply into the groove, causing it to be pinched and subjecting both the rope and sheave to severe abrasive wear. This condition can arise if new ropes are installed over old sheaves.

If the sheaves are not perfectly aligned both

OTHER EQUIPMENT

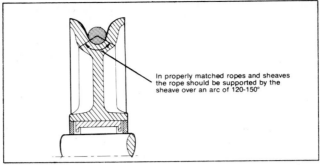

Fig. 4.9 Correct arc of support given the rope by a sheave

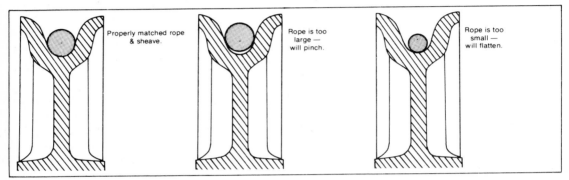

Fig. 4.10 Matching of ropes and sheaves

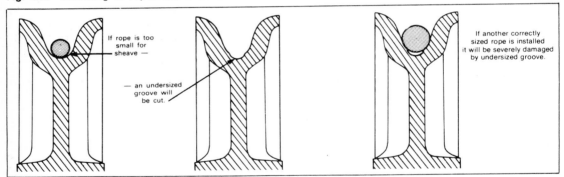

Fig. 4.11 Effect of incorrect matching of rope and sheave

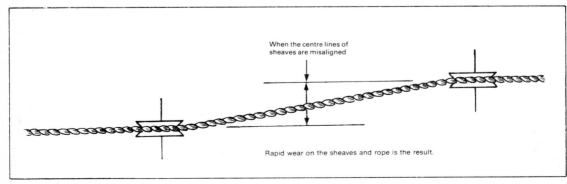

Fig. 4.12 How sheave misalignment affects rope

OTHER EQUIPMENT

TABLE 4.4

SHEAVE GROOVE TOLERANCES

Nominal rope diameter		Groove oversize			
		min.		max.	
mm	in	mm	in	mm	in
6–8	1/4–5/16	0.40	1/64	0.79	1/32
10–19	3/8–3/4	0.79	1/32	1.60	1/16
21–29	13/16–1 1/8	1.20	3/64	2.40	3/32
31–38	1 3/16–1 1/2	1.60	1/16	3.20	1/8
40–57	1 9/16–2 1/4	2.40	3/32	4.80	3/16
59 and up	2 5/16 and up	3.20	1/8	6.30	1/4

BS.302 recommends not less than 5% and approximately 7½% greater than the nominal radius of rope. See also BS.1757.

the rope and sheave flanges will be subjected to severe wear and rapid deterioration will occur. A ready indication of poor alignment is rapid wear of only one of the flanges on any given sheave. (Fig. 4.12)

One of the fastest ways to ruin a rope is to operate it over small sheaves. The excessive and repeated bending and straightening of the wires leads to premature failure from fatigue. Use the maximum possible diameter of sheave that the equipment will carry.

The sheaves should never be of smaller diameter than indicated in Table 4.3.

The depth of the sheave grooves should be at least 1½ times the rope diameter and the tapered side walls of the grooves should not make an angle of more than 26° with respect to the centre line. The flange corners should be rounded and the rims should run true about the axis of rotation. (Fig. 4.13)

The bearings should be either permanently lubricated or equipped with means for lubrication. Inadequate lubrication or a sheave that is too heavy for the load will cause the rope to slip in the sheave whenever the rope velocity changes. The momentum of the heavy sheave will cause it to continue turning after the rope has stopped. This grinding wheel action causes severe rope abrasion and will wear flat spots in the sheave that further damage the rope. (Fig. 4.14)

If the sheaves are carrying ropes that can be momentarily unloaded, as in the case of a lift line, then the sheave must be equipped with cable-keepers that prevent the unloaded rope from leaving the groove. The sheaves in lower load blocks should also be equipped with either cable-keepers or with close-fitting guards that prevent the ropes from becoming fouled when the block is lying on the ground with the ropes loose. When these guards are fitted it is important that they be removed only for the purposes of maintenance, inspection or adjustment. Failure to observe this procedure may allow a rope to jump clear and become trapped. (Fig. 4.15)

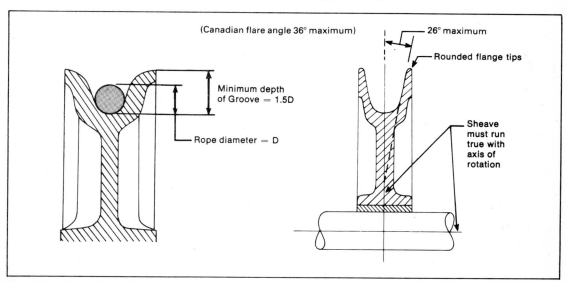

Fig. 4.13 Check that all sheaves meet these requirements

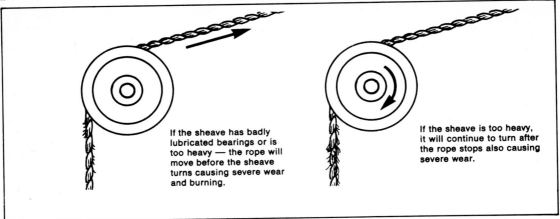

Fig. 4.14 Effects on rope life of incorrect sheaves or poor sheave maintenance

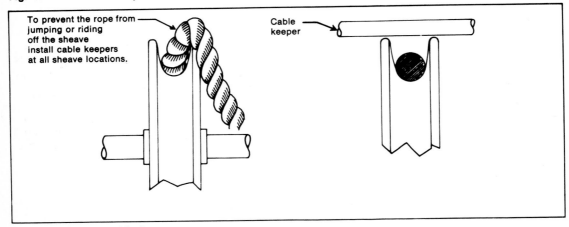

Fig. 4.15 Sheave cable keepers

Badly worn sheaves have an adverse effect on rope life and must be inspected at regular intervals. When replacement becomes necessary only equipment supplied or approved by the equipment manufacturer should be fitted. Some designs may allow for re-machining the grooves but this operation should only be undertaken in accordance with the manufacturer's instructions as there is a limit to the amount of metal which can be removed before the strength or efficiency of the component will be affected.

Inspect the sheaves carefully for any sign of cracks in the flanges. If the flange breaks off it will allow the rope to jump free with disastrous results. Even if only a small portion of the flange is broken, the rope is liable to be cut completely through by the rough edge of the break.

Never use fibre ropes on sheaves that have been used for wire rope service. And wire rope must never be used on equipment designed for fibre rope.

The groove surfaces on sheaves should be perfectly smooth. Those which have taken the imprint of the outer wires will exert a grinding action on the ropes. (Fig. 4.16)

This imprinting and scoring is caused by high contact pressures between the rope and sheave surface. If this condition is evident then the sheave must be resurfaced and the contact pressure reduced by:

(a) decreasing the load on the rope, or
(b) increasing the sheave diameter, or
(c) replacing the sheave with one made of harder material.

The calculations for radial pressures of drums in the previous section apply equally well to sheaves.

OTHER EQUIPMENT

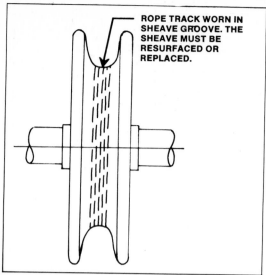

Fig. 4.16 Scored sheave groove

HOOKS

There is such a variety of hooks available for lifting operations and equipment that it is impossible to deal with all of them in detail. There are, however, several considerations that apply to all hooks and observance of them will improve the safety standard.

Make sure that all lifting hooks, excepting grab and sorting hooks, are equipped with safety catches.

Inspect all hooks frequently (Fig. 4.17). Look for wear in the saddle of the hook. Look for cracks, severe corrosion and twisting of the hook body. Be especially careful to measure the throat opening. If a hook has been overloaded or if it is beginning to weaken the throat will open. If there is any evidence of opening or distortion, destroy the hook. If you discard it without destroying it someone else may attempt to use it.

Be especially careful during the inspection to look for cracks in the saddle section and at the neck of the hook.

Buy and use only the best hooks available. These are usually forged alloy steel and generally are stamped with their rated safe working loads (SWL).

The SWL apply only when the load is applied in the saddle of the hook. If the hook is eccentrically loaded or if the load is applied anywhere between the saddle and the tip, the SWL is very much reduced. (Fig. 4.18)

Commonly-used choker hooks include:

– *Standard choker hook* designed for attachment to the end of a sling. It frees the main part of the sling of fittings. (Fig. 4.19)
– *Adjustable sliding choker hook* which prevents the hook from sliding on the sling when not loaded. A spring insert has just enough tension to hold the hook in any position on the body of the sling. The hook is easily moved by hand but will grip under load. (Fig. 4.20)
– *Dual sliding choker hook* for endless slings and conventional slings when two eyes are over main hook. (Fig. 4.21)

It is recommended that all lifting hooks, as opposed to sling hooks, should be equipped with swivels and headache balls (Fig. 4.22). Ensure that the headache ball is securely attached either to the hook or to the rope so that there is no possibility of its sliding up and down on the load line.

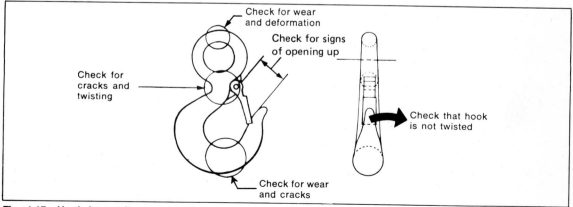

Fig. 4.17 Hook inspection areas

OTHER EQUIPMENT

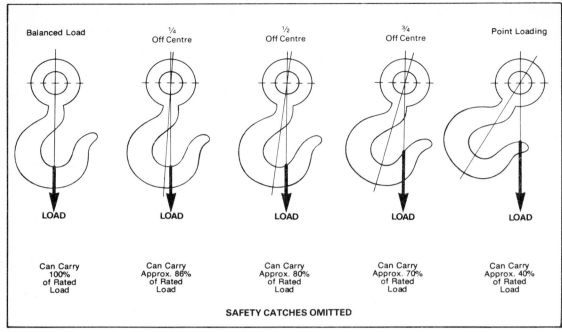

Fig. 4.18 Effect on hook capacity of eccentric loads

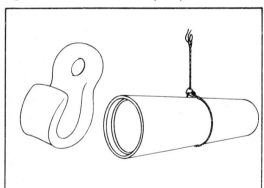

Fig. 4.19 Standard choker hook

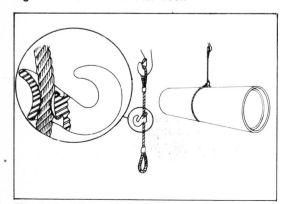

Fig. 4.20 Adjustable sliding choker hook

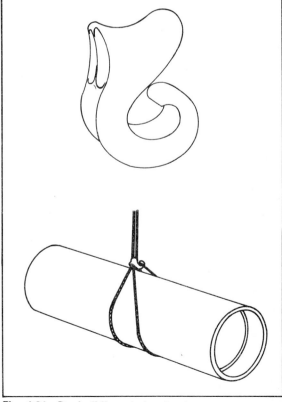

Fig. 4.21 Dual sliding choker hook

OTHER EQUIPMENT

The following tables of loads (Tables 4.5, 4.6) are included to provide an indication of what can be expected from a hook based on its throat opening. Refer to the manufacturers' ratings for specific values of specific hooks. Refer also to BS.3458 and 2903 for hook details. Table 4.6 shows a Canadian Standard.

TABLE 4.5

EYE HOOKS TO BS.3458
ALLOY STEEL

Safe Working Load = Proof Load / 2

Throat opening		Chain size		Safe working load
mm	in	mm	in	tonnes
22	7/8	6	1/4	0.76
25	1	7	9/32	0.97
32	1 1/4	10	3/8	1.68
38	1 1/2	11	7/16	2.29
43	1 11/16	13	1/2	3.05
49	1 15/16	14	9/16	3.81
54	2 1/8	16	5/8	4.72
70	2 3/4	21	13/16	8.03
76	3	22	7/8	9.30
86	3 3/8	25	1	12.19
108	4 1/4	32	1 1/4	19.05
130	5 1/8	38	1 1/2	27.43

TABLE 4.6

Canadian Standard

CHAIN SLIP HOOKS
(CLEVIS AND EYE TYPES)
FORGED ALLOY STEEL
(SAFETY FACTOR = 4)

Throat opening		For size of chain		Safe working load
mm	in	mm	in	tonnes
24	15/16	6	1/4	1.25
27	1 1/16	8	5/16	1.95
33	1 5/16	10	3/8	2.38
40	1 9/16	11	7/16	3.17
43	1 11/16	13	1/2	4.08
51	2	16	5/8	6.12
54	2 1/8	19	3/4	8.73
70	2 3/4	22	7/8	11.79
76	3	25	1	15.42

TABLE 4.7

RINGS TO BS.3458
ALLOY STEEL
Single leg sling

Stock diameter		Inside diameter		Chain size		Safe working load
mm	in	mm	in	mm	in	tonnes
16	5/8	76	3	6	1/4	0.76
19	3/4	76	3	8	5/16	1.17
22	7/8	102	4	10	3/8	1.68
25	1	102	4	13	1/2	3.05
32	1 1/4	127	5	16	5/8	4.72
38	1 1/2	152	6	19	3/4	6.86
44	1 3/4	178	7	22	7/8	9.30
51	2	203	8	25	1	12.19
64	2 1/2	267	10 1/2	32	1 1/4	19.05
76	3	305	12	38	1 1/2	27.43

TABLE 4.8

LINKS ALTERNATIVE TO RINGS TO BS.3458
ALLOY STEEL
Single leg sling

Stock diameter		Inside width		Chain size		Safe working load
mm	in	mm	in	mm	in	tonnes
41	1 5/8	108	4 1/4	21	13/16	8.03
41	1 5/8	108	4 1/4	22	7/8	9.30
44	1 3/4	121	4 3/4	24	15/16	10.67
44	1 3/4	121	4 3/4	25	1	12.19
51	2	140	5 1/2	29	1 1/8	15.39
57	2 1/4	159	6 1/4	32	1 1/4	19.05
64	2 1/2	184	7 1/4	35	1 3/8	23.01
67	2 5/8	184	7 1/4	38	1 1/2	27.43

TABLE 4.9

REEVABLE EGG LINKS TO BS.3458
ALLOY STEEL

Stock diameter		Inside length		Chain size		Safe working load
mm	in	mm	in	mm	in	tonnes
11	7/16	70	2 3/4	6	1/4	0.76
14	9/16	86	3 3/8	8	5/16	1.17
16	5/8	102	4	10	3/8	1.68
22	7/8	137	5 3/8	13	1/2	3.05
27	1 1/16	171	6 3/4	16	5/8	4.72
32	1 1/4	203	8	19	3/4	6.86
37	1 7/16	238	9 3/8	22	7/8	9.30
44	1 3/4	273	10 3/4	25	1	12.19
54	2 1/8	340	13 3/8	32	1 1/4	19.05
64	2 1/2	406	16	38	1 1/2	27.43

RINGS, LINKS, SWIVELS

These items, like all other lifting and slinging fittings, must be forged alloy steel to provide the highest degree of safety.

In the absence of manufacturers' safe working loads, Tables 4.7, 4.8, 4.9, 4.10 and 4.11 can be used to get an approximate idea of the capacities of the items.

SHACKLES

There are two types of shackle commonly used in lifting operations. They are the anchor (bow type) shackle and chain (D type) shackle both of which are available with threaded pins or round pins (Fig. 4.23)

Shackles are sized by the diameter of the steel in the bow section rather than the pin size. They should be of forged alloy steel only.

Never replace the shackle pin with a bolt, only the proper fitted pin should be used. Bolts are not intended to take the bending that is normally applied to the pin. (Fig. 4.24)

Never use a shackle if the inside width is greater than that listed in Table 4.12. All pins must be straight and all threaded pins must be completely seated. Cotter pins must be used with all round pin shackles.

Shackles worn in the crown or the pin by more than 10% of the original diameter should be destroyed. (Fig. 4.25)

Never allow a shackle to be pulled at an

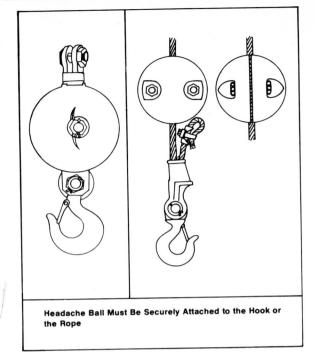

Headache Ball Must Be Securely Attached to the Hook or the Rope

Fig. 4.22 Headache ball

angle because the capacity will be tremendously reduced. Centralise whatever is being lifted on the pin by suitable washers or spacers. (Fig. 4.26)

Do not use screw pin shackles if the pin can roll under load and unscrew. (Figs. 4.27, 7.16)

Refer to Table 4.12 for the rated SWL of shackles.

TABLE 4.10

INTERMEDIATE LINKS TO BS.3458 ALLOY STEEL						Safe working load tonnes
Stock diameter		Inside width		Chain size		
mm	in	mm	in	mm	in	
8	5/16	14	9/16	6	1/4	0.76
10	3/8	18	11/16	8	5/16	1.17
13	1/2	21	13/16	10	3/8	1.68
16	5/8	27	1 1/16	13	1/2	3.05
19	3/4	35	1 3/8	16	5/8	4.72
24	15/16	41	1 5/8	19	3/4	6.86
27	1 1/16	48	1 7/8	22	7/8	9.30
30	1 3/16	54	2 1/8	25	1	12.19
38	1 1/2	68	2 11/16	32	1 1/4	19.05
46	1 13/16	81	3 3/16	38	1 1/2	27.43

TABLE 4.11

SWIVELS TO BS.4283 ALLOY STEEL		
Stock diameter		Safe working load tonnes
mm	in	
12	7/16	1.0
13	1/2	1.25
15	5/8	1.6
17	11/16	2.0
19	3/4	2.5
21	13/16	3.2
23	7/8	4.0
26	1	5.0
29	1 1/8	6.3
32	1 1/4	8.0
37	1 1/2	10.0
41	1 5/8	12.5

OTHER EQUIPMENT

TABLE 4.12

SHACKLES – DEE AND BOW TO BS.3551
ALLOY STEEL

Stock diameter				Inside width at pin				Safe working load	
Dee		Bow		Dee		Bow		Dee	Bow
mm			in	mm	in	mm	in	tonnes	
13			½	29	1⅛	22	⅞	1.12	1.12
16			⅝	32	1¼	29	1⅛	2.29	1.88
19			¾	38	1½	35	1⅜	3.05	3.05
22			⅞	44	1¾	41	1⅝	4.57	4.17
25			1	51	2	44	1¾	5.69	5.59
29			1⅛	54	2⅛	51	2	7.62	7.11
32			1¼	60	2⅜	57	2¼	9.14	8.64
35			1⅜	67	2⅝	64	2½	10.67	10.92
38			1½	70	2¾	70	2¾	14.48	12.95
41			1⅝	76	3	76	3	17.01	14.45
44			1¾	83	3¼	86	3⅜	19.81	17.53
48			1⅞	92	3⅝	92	3⅝	21.59	19.56
51			2	98	3⅞	95	3¾	24.64	22.86
58			2⅛	105	4⅛	98	3⅞	27.43	25.40
57	2¼	60	2⅜	108	4¼	111	4⅜	30.48	30.48
60	2⅜	67	2⅝	114	4½	121	4¾	35.56	35.56
67	2⅝	70	2¾	127	5	127	5	40.64	40.64
73	2⅞	76	3	140	5½	140	5½	50.80	50.80
83	3¼	90	3½	159	6¼	162	6⅜	66.04	66.04
92	3⅝	98	3⅞	178	7	178	7	81.28	81.28

TABLE 4.13

EYE BOLTS COLLAR TYPE ONLY – BS.4278
ALLOY STEEL

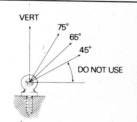

Stock diameter		Safe working load (SWL) corresponding to angle of pull				
mm	in	Vertical tonnef	75°	65°	45°	less than 45°
9	⅜	0.32				
14	9/16	1.0				
18	11/16	1.6				
23	⅞	2.5	Reduce vertical load by 45°	Reduce vertical load by 65°	Reduce vertical load by 75°	NOT RECOMMENDED
29	1⅛	4.0				
36	17/16	6.3				
46	1 13/16	10.0				
58	2 5/16	16.0				
65	2⅜	20.0				
72	2⅞	25.0				

Note: SWL for plain (collarless) eye bolts are the same as for collar type bolts under vertical load. No angular loading should be applied to plain eye bolts.

OTHER EQUIPMENT

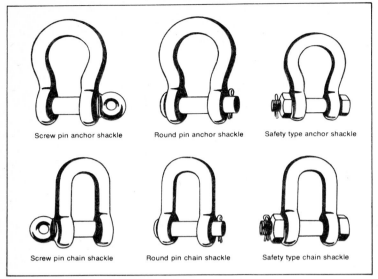

Fig. 4.23 Typical shackles

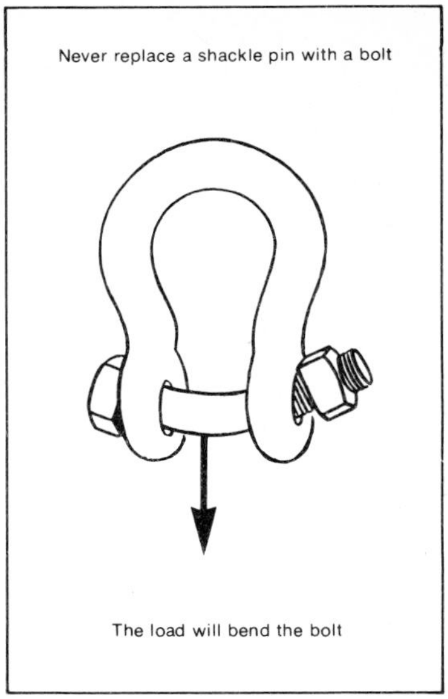

Fig. 4.24 Replacing shackle pins

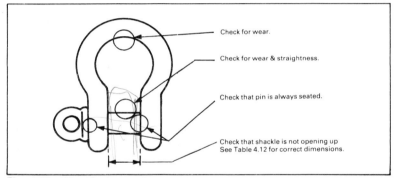

Fig. 4.25 Shackle inspection areas

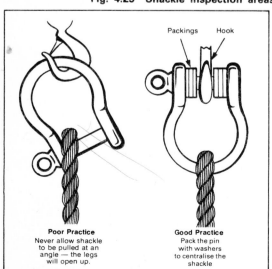

Fig. 4.26 Eccentric shackle loads

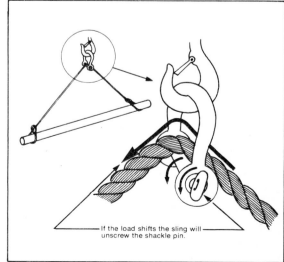

Fig. 4.27 Do not use screw pin shackles if, under load, pin can roll and unscrew

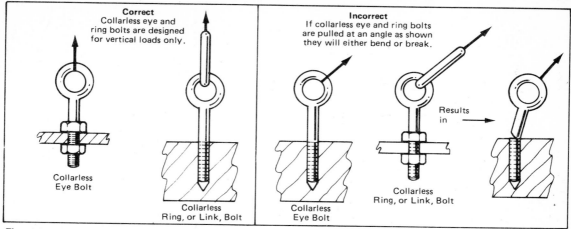

Fig. 4.28 Use of eye bolts

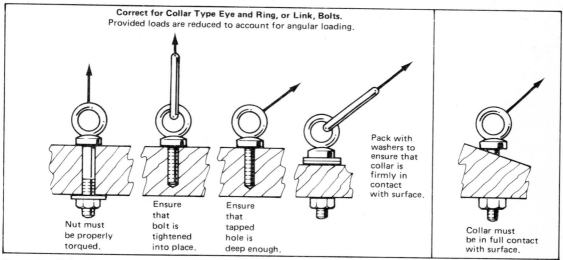

Fig. 4.29 Use of collar-type eye and ring bolts

EYE BOLTS

It is recommended that all eye bolts and ring (or link) bolts used for lifting be of forged alloy steel and equipped with collars. The plain or collarless eye bolt is good for vertical loading but as soon as it is loaded at an angle it is subjected to bending and the load it can safely carry is severely reduced. (Fig. 4.28)

Even when equipped with collars, the safe working loads of eye bolts and ring bolts are reduced with angular loading. When installed the collar must be at right angles to the axis of the hole and must contact the working surface, and the nuts must be properly torqued. Washers may have to be used to ensure that the collars are firmly in contact with the working surface. The tapped hole for screwed eye bolts (body bolts) should have a minimum depth of 1½ times the bolt diameter and must be a good fit for the screwed shank of the eye bolt. (Fig. 4.29)

Do not confuse 'dynamo' eye bolts with 'collar' eye bolts. The dynamo eye bolt has a small collar but is suitable only for axial loading. See BS.4278.

To keep the bending to a minimum, the loads should always be applied to the plane of the eye, never in the other direction. This is particularly important when bridle slings are used because an angular pull is always developed in the eye bolts, unless a spreader bar is used as part of the sling. (Fig. 4.30)

OTHER EQUIPMENT

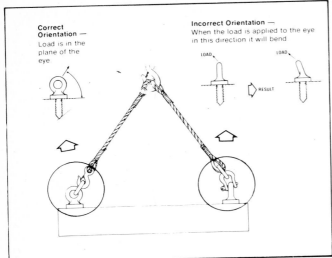

Fig. 4.30 Orientation of eye bolts

Fig. 4.31 Never insert the point of a hook into an eye bolt

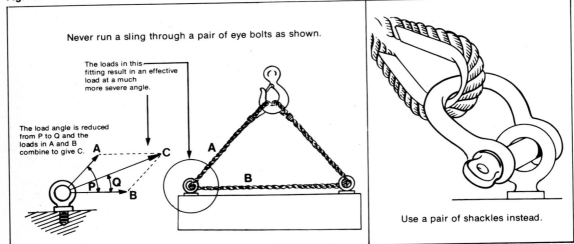

Fig. 4.32 Lifting with eye bolts

Never insert the point of a hook in an eye bolt, always use a shackle. (Fig. 4.31)

Do not use a sling reeved through an eye bolt or reeved through a pair of eye bolts. One single leg only should be attached to each eye bolt. (Fig. 4.32)

Where eye bolts cannot be kept in line with each other and at the same time tightened, thin washers or shims may be inserted under the collars to permit the eye bolts to be tightened and turned in line with each other. (Fig. 4.33)

The same precautions apply to ring bolts and the working loads are generally the same as for the eye bolts.

Refer to Table 4.13 for the rated SWL of eye bolts.

TURNBUCKLES

Turnbuckles can be supplied with eye and hook end, jaw end and stub end fittings, and any combination of these. Their rated loads depend upon the outside diameter of the threaded portion of the end fitting and on the type of end fitting. The jaw, eye and stub types are rated equally and the hook types have reduced ratings. (Fig. 4.34)

All turnbuckles used in lifting operations should be of weldless construction and fabricated from alloy steel. When they are supplied with hook end fittings the hooks should be fitted with safety catches.

If the turnbuckle is to be used in an application where vibration is present it is extremely

OTHER EQUIPMENT

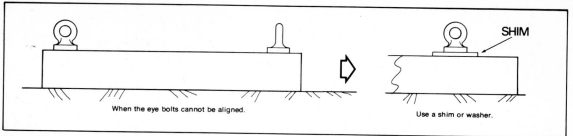

Fig. 4.33 Alignment of eye bolts

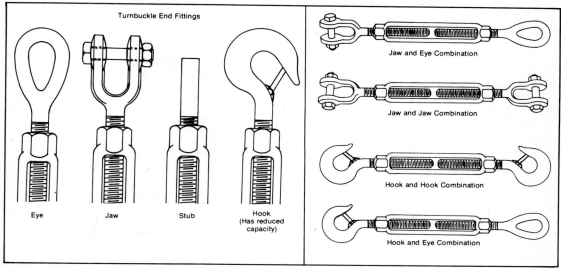

Fig. 4.34 Turnbuckles

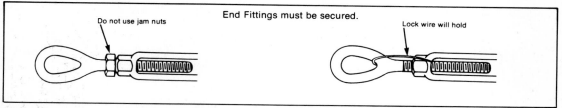

Fig. 4.35 Securing of turnbuckle end fittings

important to lock the frame to the end fittings to prevent it from turning and loosening. Lock nuts or jam nuts are not effective and add greatly to the load in the screw thread. (Fig. 4.35)

When tightening a turnbuckle, do not apply any more torque to it than you would to a bolt of equal size.

Turnbuckles should be inspected frequently for cracks in the end fittings, especially at the neck of the shank, deformed end fittings, deformed and bent rods and bodies, cracks and bends around the internally-threaded portion and signs of thread damage. (Fig. 4.36)

Refer to Table 4.14 for the (Canadian standard) rated SWL of turnbuckles.

SPREADER AND EQUALISER BEAMS

Spreader beams are usually used to support long loads during lifts. They eliminate the hazard of the load tipping, sliding or bending as well as the possibility of low sling angles and the tendency of the slings to crush the load. Equaliser beams are used to equalise

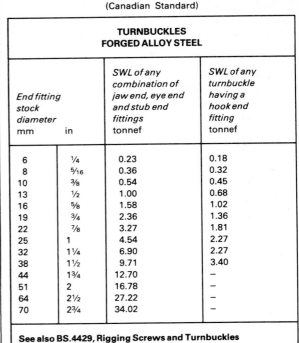

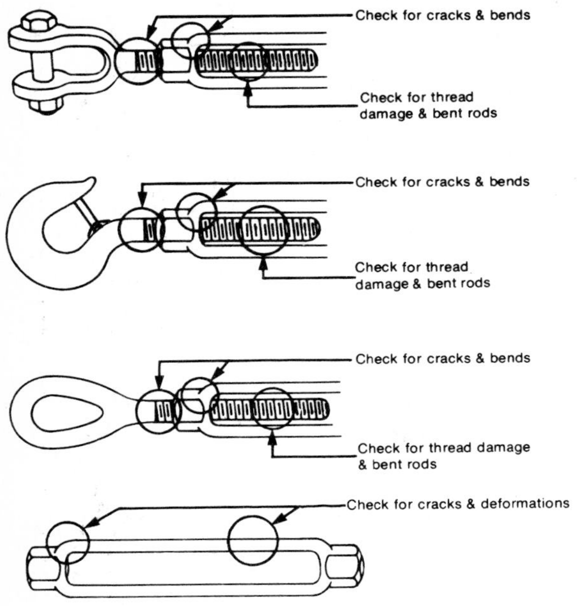

Fig. 4.36 Turnbuckle inspection areas

TABLE 4.14
(Canadian Standard)

End fitting stock diameter		SWL of any combination of jaw end, eye end and stub end fittings	SWL of any turnbuckle having a hook end fitting
mm	in	tonnef	tonnef
6	¼	0.23	0.18
8	5/16	0.36	0.32
10	⅜	0.54	0.45
13	½	1.00	0.68
16	⅝	1.58	1.02
19	¾	2.36	1.36
22	⅞	3.27	1.81
25	1	4.54	2.27
32	1¼	6.90	2.27
38	1½	9.71	3.40
44	1¾	12.70	–
51	2	16.78	–
64	2½	27.22	–
70	2¾	34.02	–

TURNBUCKLES FORGED ALLOY STEEL

See also BS.4429, Rigging Screws and Turnbuckles

the load in sling legs and/or keep equal loads on dual lift lines when making tandem lifts. (Fig. 4.37)

Both types of beam are normally fabricated to suit a specific application. If a beam is to be used which has not been designed for the application, make sure it has adequate width, depth and length, and is of suitable material.

The capacity of beams with multiple attachment points depends upon the distance between the points. For example, if the distance between the attachment points is doubled, the capacity of the beam is halved.

BLOCKS

The lifting and moving of heavy objects or objects weighing more than the safe working load of the rope being used calls for multiplication of the pulling force or changing the direction of pull, or both. Blocks make this possible.

The blocks used in construction rigging can run from the custom-designed 400-tonne capacity unit weighing 20 tonnes through all types of crane and hook blocks, wire rope blocks and snatch blocks down to the simplest of tackle blocks.

The essential parts of any block are the shell, the sheaves, the centre pin, the straps and the connections (Fig. 4.38). They can be equipped with a number of end fittings including hooks, wedge sockets, and swivels and clevis shackles of all types as well as combinations of these items. The blocks may also be equipped with a 'becket', the term used for a rope-end anchorage point on a block.

The shell provides protection for the sheave or sheaves and acts as a guide to keep the rope in place in the sheave groove. Wood shells add no strength to the block but merely serve as an enclosure for the sheaves, and they are used for fibre ropes only. Steel shells are fitted on all blocks used with wire rope. They add strength, rigidity and protection for the block.

The straps and side or cheek plates perform the important structural function of transmitting the sheave load to the connections and add rigidity to the block.

The sheaves transmit the load imposed by the rope to the centre pin, straps and connections. On fibre rope blocks the sheaves are

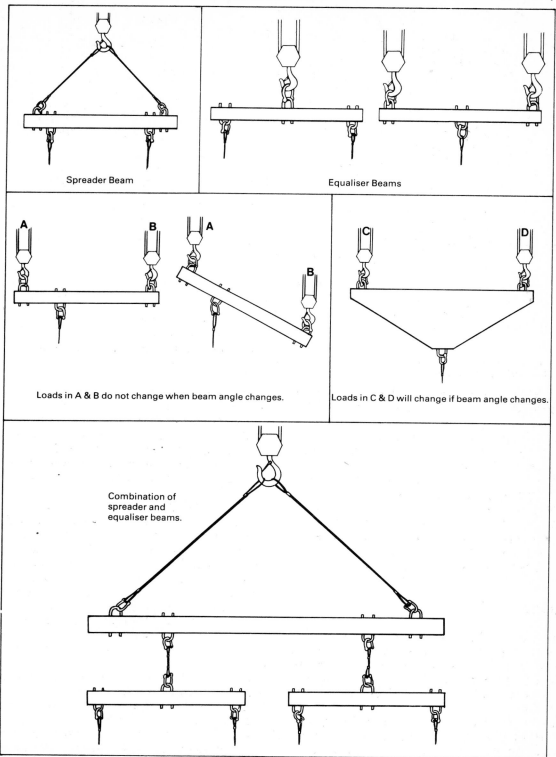

Fig. 4.37 Spreader and equaliser beams

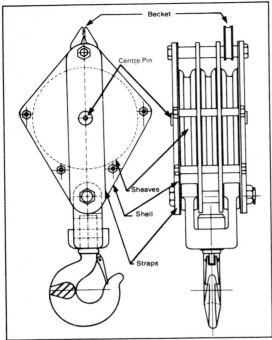

Fig. 4.38 Typical wire rope block

normally fabricated from cast iron but on wire rope blocks they should be cast steel because of its greater strength, hardness and abrasion resistance. Never run fibre ropes over sheaves that have been used with wire ropes – rapid rope damage will result. Fibre rope sheaves must not be used with wire rope because their diameters are too small.

In *plain bore sheaves* (Fig. 4.39A) the cast iron of the sheave provides its own bearing. They are recommended for light, intermittent service only and they must be frequently oiled to reduce friction and avoid excessive wear.

Roller bushed sheaves (Fig. 4.39B) are made with unground rollers and without races, and are recommended for light, intermittent service only. They must be lubricated periodically with a heavy grease.

Self-lubricating bronze bearings are recommended where service requirements are severe or where the block cannot be subjected to high speed or continuous operation. The plug type of bearing should not be lubricated or oiled. Oil dissolves and washes away the graphite-wax mixture, destroying the

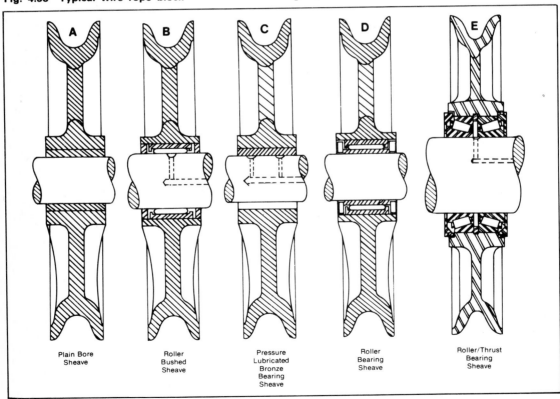

Fig. 4.39 Sheave bearing configurations

OTHER EQUIPMENT

bearings' self-lubricating characteristics. Oil is not harmful to the graphite-impregnated type of bearing.

Pressure lubricated bronze bearings (Fig. 4.39C) are recommended where loads are extra heavy and continuous. Frequent periodic lubrication is required.

Roller bearing sheaves (Fig. 4.39D) are made with ground rollers and full races and are recommended for medium duty and high-speed operation.

Precision anti-friction bearings (Fig. 4.39E) capable of sustaining both radial and thrust loads are particularly suited to continuous high operating speeds and heavy loads. They combine long life with minimum maintenance.

The centre pin, sometimes called the sheave pin, transmits the sheave load to the strap. It should be prevented or restrained from turning. (Fig. 4.40)

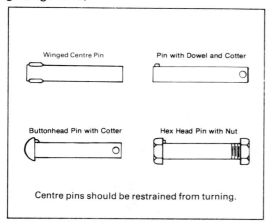

Fig. 4.40 Block centre pins

All fittings on the blocks should be forged alloy steel. Shackles and eyes are inherently stronger than hooks; blocks equipped with them are rated at higher working loads. They are recommended for block connections where the block is to be mounted permanently, where accidental disengagement of the hook connection would be dangerous or where higher strength ratings are needed. Shackle connections are particularly recommended for standing blocks which must support not only the load, but also the lifting strain of the lead line.

Blocks usually take their name from the purpose for which they are used, the position they occupy or from a particular shape or type of construction. They can be designated according to the number of sheaves they contain as single, double, triple, etc., or in accordance with the shell shape as diamond pattern, oval pattern, etc.

For the purposes of this manual, however, the most commonly-used types of block are designated as:

(a) *Crane and hook blocks.* (Fig. 4.41)
(b) *Wire rope blocks.* (Fig. 4.42)
(c) *Snatch blocks.* (Fig. 4.43)
(d) *Tackle blocks.* (Fig. 4.44)

The basic differences between these types relate to their construction and are evident from the illustrations. *Crane and hook blocks* are equipped with heavy iron cheek weights whereas the *wire rope blocks* are normally much lighter and are equipped with cheek straps that provide strength between the end attachments and sheave centre pins. Both types are well suited to high-speed applications and heavy loads. However, the wire rope blocks are not intended to withstand the heavy service and abuse expected of the crane and hook blocks.

A *snatch block* can be a single- or multi-sheave block that opens on one side to permit a rope to the slipped over the sheave thus eliminating the need for it to be threaded through the block. They are available in many configurations for wire rope, manila rope and synthetic ropes, and can be purchased with hook, shackle, eye and swivel end fittings.

Snatch blocks are normally used when it is necessary to change the direction of the pull on a line. In cases such as this, the stress on the snatch block varies tremendously with the

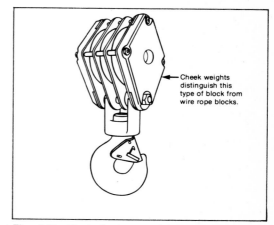

Fig. 4.41 Typical crane and hook block

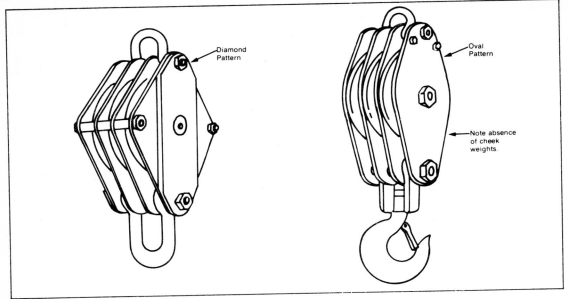

Fig. 4.42 Typical wire rope blocks

angle between the lead and load lines. When the two lines are parallel, 500 kg (1000 lb) on the lead line results in a load of 1000 kg (2000 lb) on the block, hook and whatever it is hooked onto. As the angle between the lines increases, the stress on the block and hook is reduced, as illustrated in Fig. 4.45.

To determine the load on the block, hook and anchorage point, multiply the pull on the lead line or the weight of the load being lifted by a suitable factor from Table 4.15 and add 10% for sheave friction.

Tackle blocks are used in conjunction with fibre ropes, both natural and synthetic fibre, and are similar to wire rope blocks except for their much lighter weight and capacity. They are available with either wood or metal shells and either with or without cheek straps, depending on their capacity.

The sheaves in the blocks should be of the correct size for the rope being used and must be free and well lubricated.

With the exception of snatch blocks, the other types of blocks (i.e. crane and hook, wire rope, tackle) can be classified according to their position in a reeved system. A travelling or fall block is a block attached to the load being lifted or moved and moves with the load. A standing block is a block fixed to a stationary object acting as the reaction point of the loading. (Fig. 4.46)

Many factors govern the selection and use of tackle blocks. Trouble will result from overloading, excess friction, angle of pull, condition of rope, sudden application of load, or lack of lubricant. The actual weight of the load to be moved does not necessarily determine the stress on the blocks. Obstruction to the free movement of the load, twisted ropes due to improper reeving or rigging, or improper angle of the tackle in relation to the load should be avoided. Moving heavy loads

TABLE 4.15

MULTIPLICATION FACTORS FOR SNATCH BLOCK LOADS	
10°	1.99
20°	1.97
30°	1.93
40°	1.87
50°	1.81
60°	1.73
70°	1.64
80°	1.53
90°	1.41
100°	1.29
110°	1.15
120°	1.00
130°	.84
140°	.68
150°	.52
160°	.35
170°	.17
180°	.00

OTHER EQUIPMENT

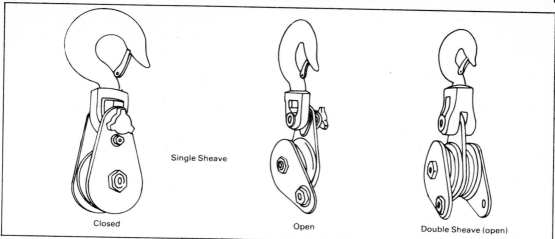

Fig. 4.43 Typical snatch blocks

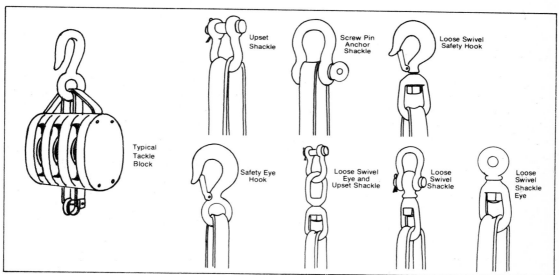

Fig. 4.44 Tackle block fittings

over rough ground, on an incline, without rollers or with too small rollers, can introduce severe stresses. A load suspended on two sets of blocks should be evenly distributed or one set will be subjected to more than its calculated share of the load. It is possible through careless preparation to have any or all of these conditions present to such a degree that the load on a set of blocks will greatly exceed the actual weight of the load itself.

In selecting blocks the governing consideration should be the load to be handled, rather than the diameter or the strength of the rope they will carry. In multi-sheave blocks the load is distributed among several parts of the rope, whereas the hooks or shackles on the blocks have to carry the entire load. It is not practical to make double blocks twice as strong and triple blocks three times as strong as single blocks, since they would be too heavy and needlessly expensive for general use.

A suitable working load is not the greatest load that blocks can lift safely, but a load which blocks can handle efficiently until worn out. Increasing the number of sheaves multiplies power at the sacrifice of lifting speed.

When blocks are used for heavy loads and

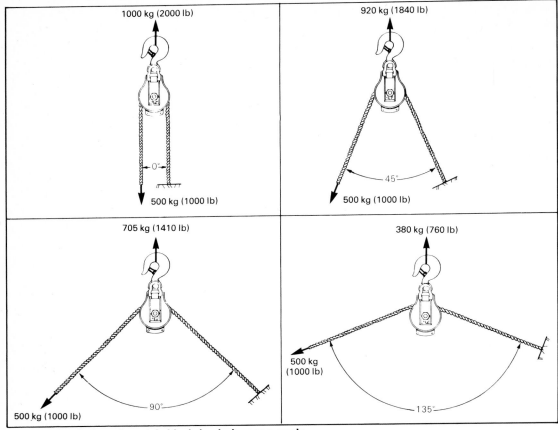

Fig. 4.45 Variation of snatch block loads by rope angle

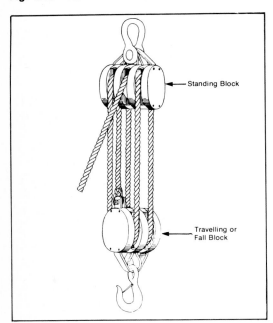

Fig. 4.46 Tackle block nomenclature

fast lifting, we strongly recommend roller or bronze bearings in the sheaves. For wire rope blocks, cast steel sheaves are recommended.

Regardless of the type or capacity of block selected, ensure that the sheaves are of the maximum possible diameter for the size of rope being used and that the blocks are clearly and permanently marked with their weight, where significant, and their rated capacity.

Remember also that the anchorage point for the tackle or blocks must be able to carry the total weight of the load plus the weight of the blocks, plus the pull exerted on the lead line.

Like any piece of equipment, all blocks require some attention. They cannot be abused and neglected without suffering loss of effectiveness and life. Keep them clean. Remove the sheaves occasionally and clean and oil the centre pins and inspect the sheaves for wear. When not in use store the blocks in a dry place.

The following inspection and examination recommendations should be observed.

- Check the blocks for excessive wear on the beckets, end connections, sheave bearings and centre pins.
- Ensure that the sheave grooves are smooth. If a wire rope sheave shows the imprint of the rope, excessive rope wear will occur.
- Look for signs of overloading; elongated links, eyes or shackles; bent shackle, link or centre pins; enlarged hook throats. If such conditions are found, the block should be replaced.
- Check the sheaves for free rotation.
- Ensure that all rope keepers are in place.
- Check the clearance between sheaves and cheek and partition plates. It should be small enough to ensure that there is no danger of the rope slipping between them.

CHAPTER 5

Reeving

The selection of blocks and their reeving for a particular application involves not only consideration of the working load, connections, sheaves, pins and other structural details, but also the mechanical advantage required. Both the size of the block, the rope size and the number of sheaves per block must be determined in relation to the load to be lifted and the pull which can be applied to the lead line.

Blocks, when reeved, become a machine, which by definition is a device by which forces may be advantageously applied to do work. A multi-part reeved system is a device or machine by which the lead line pull is multiplied to lift the load, thus gaining a mechanical advantage (purchase).

Since this system is a machine, like all machines it is not 100% efficient because of friction. As soon as the load being lifted begins to move, part of the applied force to the lead line is lost in overcoming the friction of turning the sheave and bending the rope. Hence, the lead line pull must be increased to make up for this friction loss in the tackle. (Fig. 5.1)

This relationship is called the 'efficiency' and is usually expressed as a percentage. For a greater number of sheaves and parts of line, the mechanical advantage is greater, but the friction loss is also greater and the efficiency lower. Conversely, the lesser the number of sheaves and parts of line, the lower the friction loss and the higher the efficiency, but the mechanical advantage is less.

The mechanical advantage of any multi-part reeved system is always equal to the number of parts of line supporting the running (or travelling) block and the load. The lead line should not be included. (Fig. 5.2)

The number of parts of line in a reeved system is determined as follows:

– Imagine that you are going to cut all the ropes just above the load (exclude the lead line).

TABLE 5.1

FACTORS TO ACCOUNT FOR SHEAVE FRICTION LOADS		
FRICTION FORCE = 10% of sheave load (Typical for plain bore sheaves and poorly maintained bronze bushing sheaves)		
Number of parts of line N	Multiplication factor F	Ratio R ($R = N/F$) = actual mechanical advantage
1	1.10	.91
2	1.21	1.65
3	1.33	2.26
4	1.46	2.74
5	1.61	3.11
6	1.77	3.39
7	1.94	3.61
8	2.14	3.74
9	2.36	3.81
10	2.60	3.85
11	2.85	3.86
12	3.14	3.82
13	3.45	3.77
14	3.80	3.68
15	4.18	3.59
16	4.60	3.48
17	5.06	3.36
18	5.56	3.23
19	6.12	3.11
20	6.73	2.97

REEVING

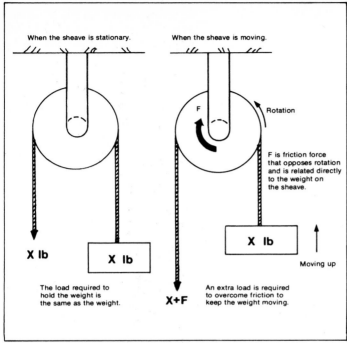

Fig. 5.1 Effect of sheave friction

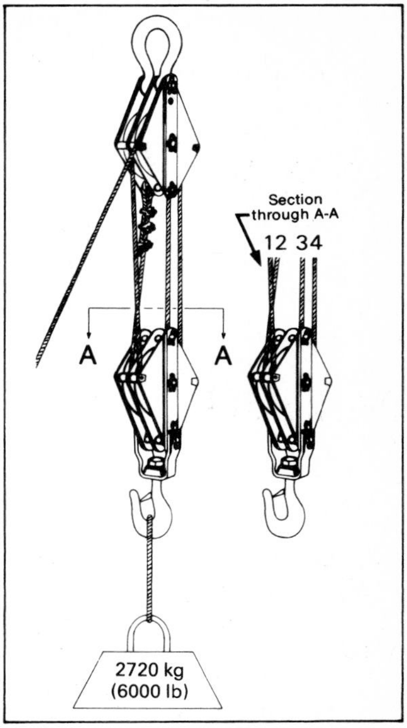

Fig. 5.2 Determination of parts of line

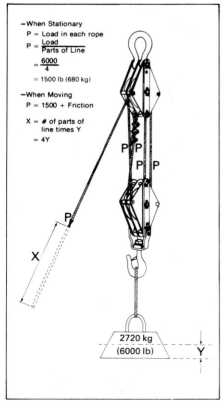

Fig. 5.3 Effect of mechanical advantage

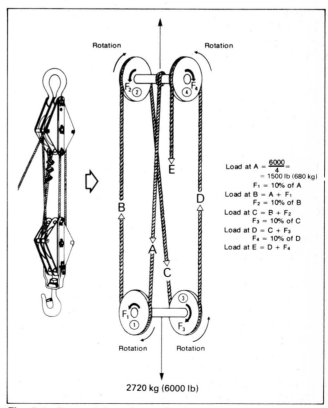

Fig. 5.4 Determining sheave loads

- Count the cut ends.
- Your answer gives you the number of parts of line which, in turn, is equal to the mechanical advantage of the system.

TABLE 5.2

FACTORS TO ACCOUNT FOR SHEAVE FRICTION LOADS
FRICTION FORCE = 3% of Sheave Load
(Typical for bronze bushing sheaves and stiff roller bearing sheaves)

Number of parts of line N	Multiplication factor F	Ratio R (R = N/F) = Actual mechanical advantage
1	1.03	0.97
2	1.06	1.89
3	1.09	2.75
4	1.12	3.57
5	1.15	4.34
6	1.18	5.08
7	1.22	5.73
8	1.26	6.35
9	1.30	6.92
10	1.34	7.46
11	1.38	7.97
12	1.42	8.45
13	1.46	8.90
14	1.50	9.33
15	1.55	9.68
16	1.60	10.00
17	1.65	10.30
18	1.70	10.59
19	1.75	10.86
20	1.80	11.11

The increase in load-carrying ability, however, is accompanied by a corresponding decrease in speed of travel. The lead line must move four times as fast as the load for a mechanical advantage of 4. The travel of the lead line, divided by the travel of the load is also a measure of the mechanical advantage. (Fig. 5.3)

If this 3000 kg (6000 lb) load is not only supported but is also raised, then it is necessary to exert more than 750 kg (1500 lb) on the lead line because of the extra forces necessary to overcome friction in the sheave bearings and the bending resistance of the rope as it passes over the sheaves. The mechanical advantage is decreased by the friction.

For practical purposes the sheave friction losses on well-maintained sheaves are approximately 10% for each sheave in manila rope blocks (plain bore sheaves) (Table 5.1), about 3–5% for each sheave having bronze bushings (Table 5.2), and approximately 1½%–3% for sheaves having roller bearings (Table 5.3).

When the load is lifted, each sheave introduces a friction force equal to either 10%, 3–5% or 1½–3% *of the load it is carrying* (depending on the sheave bearings and how easily they rotate).

If the previously-illustrated example was for sheaves having plain bores, (10% friction), the lead line load could be calculated as follows:

(See Fig. 5.4)

Load at A = 1500 lb (680 kg)

Load at B = 1500 lb + 10% of 1500
          (friction at sheave 1)
        = 1650 lb (748 kg)

Load at C = 1650 lb + 10% of 1650
          (friction at sheave 2)
        = 1815 lb (823 kg)

Load at D = 1815 lb + 10% of 1815
          (friction at sheave 3)
        = 1996 lb (905 kg)

Load at E = 1996 lb + 10% of 1996
          (friction at sheave 4)
        = 2196 lb (995 kg)
        = Lead Line Load

If you simply multiply the number of sheaves by 10% and then increase the static load on the lead line by that amount, the calculated lead line loads will be significantly *lower* than the actual loads.

The assessment and determination of friction loads requires a progressive type of line-by-line calculation.

To simplify matters, these calculations can be replaced with one calculation and the use of tables of multiplication factors and ratios.

$$\text{Lead Line Load} = \frac{\text{Load to be Lifted}}{\text{Parts of Line}} \times F$$

OR

$$\text{Lead Line Load} = \frac{\text{Load to be Lifted}}{R}$$

TABLE 5.3

FACTORS TO ACCOUNT FOR SHEAVE FRICTION LOADS
FRICTION FORCE = 1½% of sheave load
(Typical for good roller bearing sheaves)

Number of parts of line N	Multiplication factor F	Ratio R ($R = N/F$) = actual mechanical advantage
1	1.015	0.985
2	1.03	1.942
3	1.045	2.87
4	1.061	3.77
5	1.077	4.64
6	1.093	5.489
7	1.109	6.312
8	1.126	7.105
9	1.143	7.874
10	1.160	8.620
11	1.177	9.345
12	1.195	10.041
13	1.213	10.717
14	1.231	11.373
15	1.249	12.01
16	1.268	12.618
17	1.287	13.209
18	1.306	13.783
19	1.325	14.34
20	1.345	14.87

Values of F and R are provided in Tables 5.1, 5.2 or 5.3, depending on the sheaves.

Fig. 5.5 is a graph of the R values in the Tables, and the number of parts of line. This graph shows the effect of the number of parts of line on the actual mechanical advantage which considers friction. For example, for a system having plain bore sheaves and four parts of line, the graph shows that the actual mechanical advantage is only 2.74, not 4 (Table 5.1). This is because the sheave friction has reduced the advantage of the extra parts of line. Increasing the number of parts of line on blocks having plain bore sheaves can work to a disadvantage beyond 10 or 11 parts of line and does not really help matters much beyond six or seven parts of line. This, of course, is due to the cumulative effect of sheave friction. The point is rapidly reached where the largest portion of the input load is used to overcome friction rather than move the load.

The situation is considerably better on blocks with bronze bushings and roller bearings but they too, as shown in the graph, reach a point where the addition of more sheaves and parts of line does not help matters much. Remember, too, that a poorly-maintained or insufficiently lubricated sheave with either bronze bushings or roller bearings can have friction loads much higher than the values quoted.

The following are three very important calculations that every crane driver and slinger should know:

(a) How to determine the maximum load that can be lifted with a given reeving arrangement. (Table 5.1)
(b) How to determine the line pull (or rope size) needed when the load weight and number of parts of line are established. (Table 5.2)
(c) How to determine the number of parts of line required to make a lift. (Table 5.3)

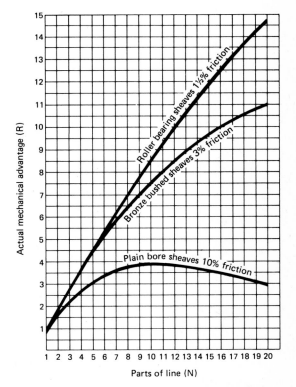

Fig. 5.5 True mechanical advantage against parts of line

The formulae needed to make these calculations are as follows:

(a) To determine the maximum load that can be lifted with a given reeving arrangement:

- Determine the number of parts of line.
- Find out what kind of sheave bearings are being used and how easily they rotate.
- Determine the maximum SWL of the rope and the maximum pulling load that you can or want to apply to the lead line. Use the smallest of the two values in the calculation.
- Determine R from appropriate table opposite the number of parts of line.
- Calculate as follows:

MAXIMUM LOAD = ROPE SWL (OR LEAD LINE PULL) × R

Example: Tackle blocks have four parts of line and plain bore sheaves. 16 mm (5/8 in) diameter manila rope is used (SWL = 400 kg (880 lb)) but the maximum load that can be applied to the lead line is only 180 kg (400 lb).

From Table 5.1 for plain bore sheaves, or from the graph (Fig. 5.5), opposite four parts of line, find that R = 2.74.

MAX. LOAD = 180 kg (400 lb) × 2.74
         = 498 kg (1095 lb)

This means that 498 kg (1095 lb) is the maximum load that can be lifted.

(b) To determine the line pull (or rope size) needed when the load weight and number of parts of line are established:

- Determine the type of sheave bearings.
- Determine the load weight.
- Determine the number of parts of line.
- Divide the load weight by the R factor that applies to the number of parts of line and the type of sheave bearings.

$$\frac{\text{LOAD WEIGHT}}{R} = \text{LINE PULL}$$

Example: A hoist is rigged with six parts of line and the sheaves have bronze bushings. The load to be lifted weighs 12 200 kg (27 000 lb).

From Table 5.2 or Fig. 5.5 for bronze bearing sheaves and opposite six parts of line, read R = 5.08.

Therefore, the required line pull is:

$$\text{LINE PULL} = \frac{\text{LOAD WEIGHT}}{R}$$

$$= \frac{12\ 200\ \text{kg}\ (27\ 000\ \text{lb})}{5.08}$$

$$= 2400\ \text{kg}\ (5300\ \text{lb})$$

This means that the lift winch must be able to exert a line pull of 2400 kg (5300 lb) if it is to lift the load. Ensure that the required line pull does not exceed the rated SWL of the wire rope.

These recommendations and calculations apply not only to blocks but also to individual sheaves, idlers and snatch blocks. Every sheave that a rope is in contact with adds frictional loads to that rope. As has been demonstrated, when the friction is high and there are many sheaves these loads can combine to the point where they actually *exceed* the loading due to the weight of the object being lifted.

It is important that blocks and reeved systems be of sufficient size and capacity to carry the loads to which they will be subjected. If the hook on a set of blocks shows signs of opening up then that is an indication that the blocks are being overloaded since the hook will usually begin to bend at approximately 70% of the maximum block strength. The best method of avoiding overloads is to know exactly how reeved systems work, use the previously-

(c) To determine the number of parts of line required to make a lift:

- Determine the maximum SWL of the rope and the maximum pulling load that you can or want to apply to the lead line. Use the smaller of the two values in the following calculation.
- Determine the weight of the load to be lifted.
- Calculate as follows:

$$\frac{\text{LOAD TO BE LIFTED}}{\text{ROPE SWL (OR DESIRED LEAD LINE PULL)}} = R$$

- Opposite R in the Tables read the number of parts of line, or make the determination from the graphs.

Example: Rope block sheaves have roller bearings and are well maintained. Crane to be rigged for capacity of 50 tonnes. Rope's SWL is equal to 8.6 tonnes.

$$\frac{\text{LOAD}}{\text{SWL}} = \frac{50}{8.6} = 5.8 = R$$

By reading down the R column in Table 5.3 for roller bearing sheaves we see that our R of 5.8 falls between the 5.49 and 6.31 figures — always go to the next highest figure, which in this case is 6.31. Opposite the 6.31 we see that this corresponds to seven parts of line. Therefore, this machine requires seven parts of line to make its lifts.

---

mentioned calculations and appreciate just how much effect friction has on the total load.

If the blocks you are going to use have *equal* numbers of sheaves then the dead end of the rope should be fastened to the becket of the standing (upper) block. If they have *unequal* numbers of sheaves then the rope should be fastened to the becket of the block with the least number of sheaves (Fig. 5.6). Make sure the rope termination at the becket is secure and develops as much strength as the rope itself. Refer to the chapters on wire rope and fibre rope to determine the best anchorage methods.

In reeving a pair of tackle blocks one of which has more than two sheaves, the lifting rope should lead from one of the centre sheaves of the upper block.

When reeved this way, the lifting strain falls on the centre of the blocks, thus preventing them from toppling and twisting and causing injury to the rope by cutting across the edges of the block shell.

It is good practice to use a shackle block as the upper one of a pair and a hook block as the lower one. A shackle is much stronger than a hook of the same size. Also, the strain on the upper block is much greater than on the lower one. The lower block supports only the load whereas the upper block carries the load as well as the lifting strain. A hook is more convenient on the lower block because it is easier to attach to or detach from the load.

When making up a set of falls or reeving a system having five or more parts of line, avoid lacing the blocks because they have a tendency to tilt the travelling block when running unloaded. (Fig. 5.7) This effect causes excessive wear and can damage the sheaves and rope. Whenever possible, right-angle reeving should be used. The blocks should be identical and square in relation to each other.

When reeving any kind of rope block it is extremely important that the dead end, or becket connection, be properly made and attached. On wire rope blocks use either a wedge socket or cable clips for the becket. On fibre rope blocks use the becket hitch (Fig. 5.8) to secure the dead end.

Two-Part Falls: (Fig. 5.9) (two single-sheave blocks) — Feed the lead line over the sheave of the stationary block, down behind the travelling block, through the sheave and up to the becket on the stationary block. Secure to the becket.

Three-Part Falls: (Fig. 5.10) (two-sheave block plus single-sheave block) — Feed the lead line through the front of the stationary block at sheave A, then feed it down behind the travelling block and through sheave C. Bring it up in front of the stationary block, pass it through sheave B and down to the becket on the travelling block. Secure to the becket.

REEVING

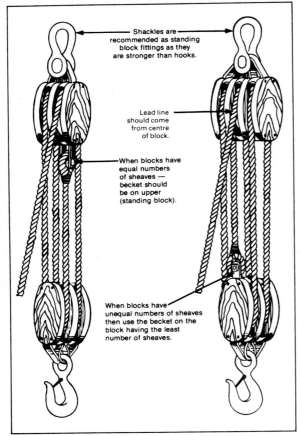

Fig. 5.6 Block orientation in a reeved system

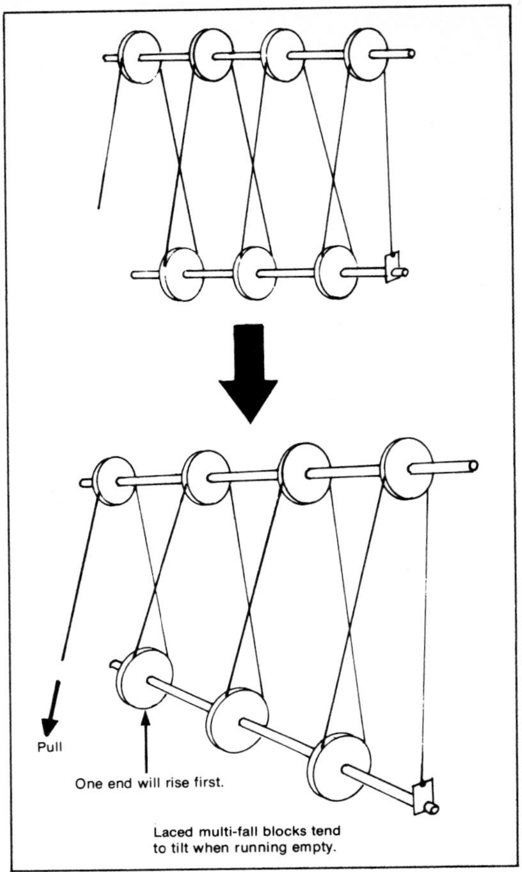

Fig. 5.7 Effect of lacing rather than reeving blocks

Fig. 5.8 Becket hitch

REEVING

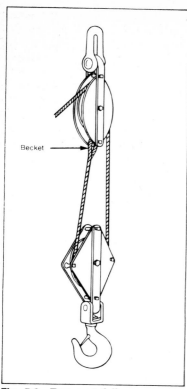

Fig. 5.9 Two-part falls

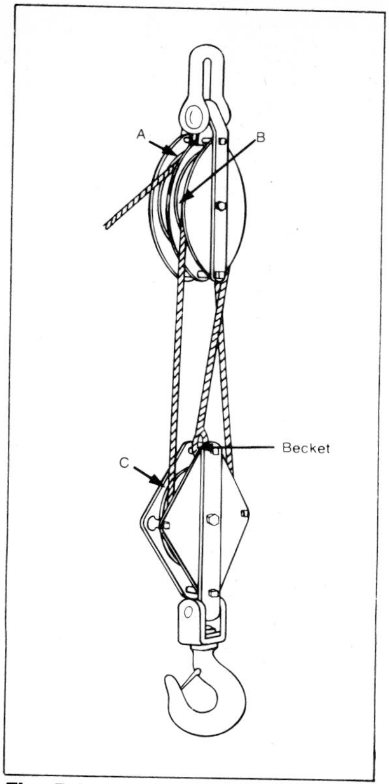

Fig. 5.10 Three-part falls

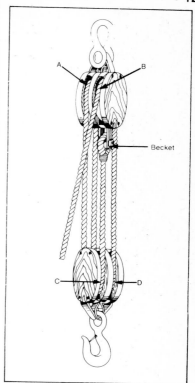

Fig. 5.11 Four-part falls

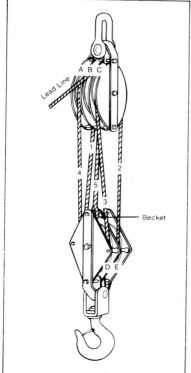

Fig. 5.12 Five-part falls

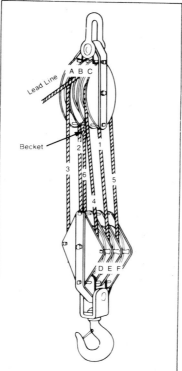

Fig. 5.13 Six-part falls

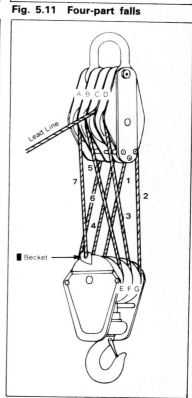

Fig. 5.14 Seven-part falls

Four-Part Falls: (Fig. 5.11) (pair of two-sheave blocks) – Arrange the blocks at right angles to each other. Feed the lead line through the front of the stationary block at sheave B, then feed it down in front of the travelling block and through sheave D. Bring it up in behind the stationary block and feed it through sheave A. Take the lead line down in behind the travelling block, pass it through sheave C and bring it up to the becket on the stationary block. Secure to the becket.

Five-Part Falls: (Fig. 5.12) (two-sheave block plus three-sheave block) – Feed the lead line through the front of the stationary block at sheave B, then down Behind the travelling block, and through sheave E. Bring up behind the stationary block and through sheave C, down in front of travelling block and through at sheave D. Then bring it up in front of stationary block and through at sheave A, down to the travelling block and secure to the becket.

Six-Part Falls: (Fig. 5.13) (pair of three-sheave blocks) – Feed the lead line through the front of the stationary block at sheave B, then down in front of the travelling block and through at sheave E. Bring it up behind the stationary block and through at sheave A, down behind the travelling block and through at sheave D. Then bring it up in front of the stationary block and through at sheave C, down in front of the travelling block, and through at sheave F. Finally, bring it up to the stationary block and secure to the becket.

Seven-Part Falls: (Fig. 5.14) (three-sheave block plus four-sheave block) – Feed the lead line through the front of the stationary block (four-sheave) at sheave C, down in front of the travelling block and through at sheave F. Bring up behind the stationary block and through at sheave A, down behind the travelling block and through at sheave E. Then up in front of the stationary block and through at sheave D, down in front of the travelling block and through at sheave G. Finally, up behind the stationary block and through at sheave B, then down to the travelling block and secure to the becket.

Eight-Part Falls: (Fig. 5.15) (pair of four-sheave blocks) – Feed the lead line through the front of the stationary block at sheave C, down in front of the travelling block and through at sheave G. Next bring it up behind the stationary block and through at sheave A, down behind the travelling block and through at sheave E. Then up in front of the stationary block and through at sheave D, down in front of the travelling block and through at sheave H, up behind the stationary block and through at sheave B. Finally, bring it down behind the travelling block and through at sheave F, then up to the stationary block and secure to the becket.

If the blocks have more sheaves than are needed, ensure that they are symmetrically reeved with the rope as equally distributed across the block as possible to equalise the loads and prevent toppling. (Fig. 5.16)

If an auxiliary hook or line is run over the jib head or over a standing block, the main hook or line is generally reeved to one side of the jib head or block. This eccentric condition introduces severe torsional stresses in the jib and can cause blocks to topple. When approaching capacity loads, the blocks and jib head must be reeved in accordance with Fig. 5.17.

All tackles and reeved systems, particularly long new ones, are liable to twist primarily because of lay of the rope, not the method of reeving. If this happens, the power required to lift increases and sheave damage and rope wear occur. Unfortunately, there is little that can be done other than use two-in-one braided fibre or non-rotating wire rope.

If the system is rigged with only a single-rope fall there is always a tendency for the rope and hook to spin after the tension is removed from the lifting rope. To avoid being hit by the hook or load stay away from them until they are set down.

Another consideration in the use of reeved systems is that all sheaves in a set of blocks revolve at different rates of speed. The sheaves nearest the lead line rotate at the highest rates and consequently need more lubricant than the others and also wear out more rapidly.

REEVING

129

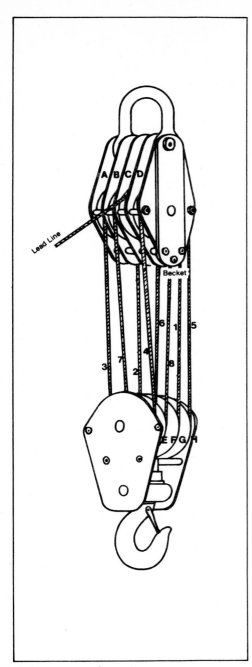

Fig. 5.15 Eight-part falls

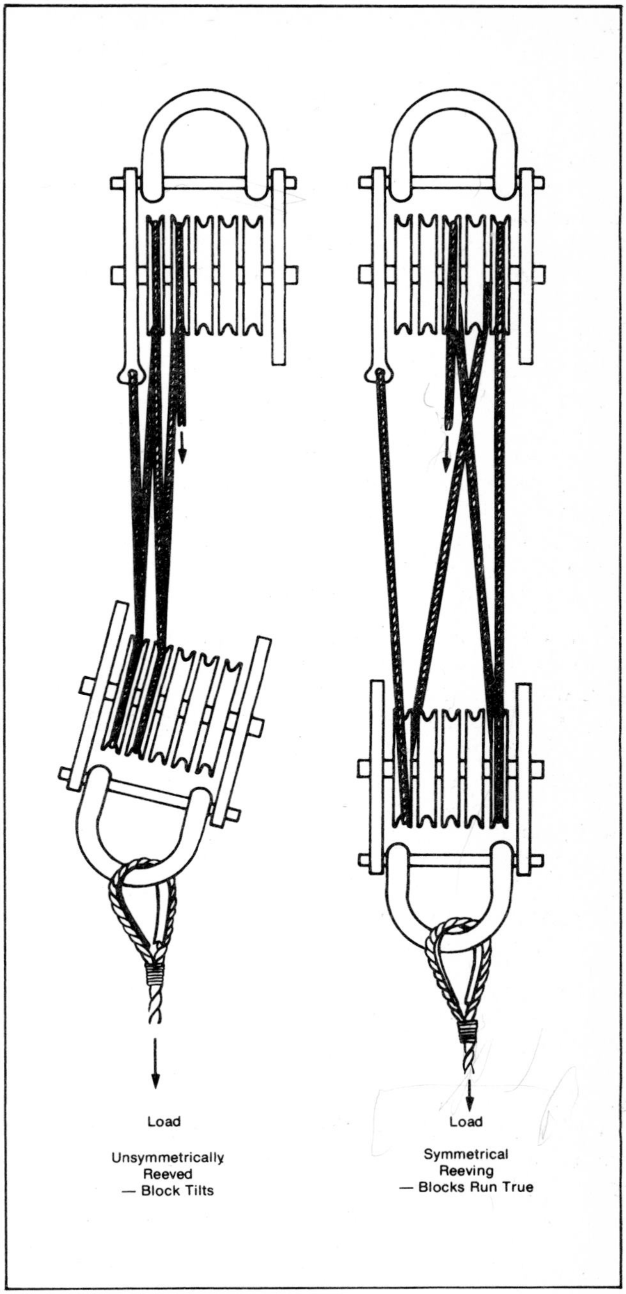

Fig. 5.16 Symmetrical reeving

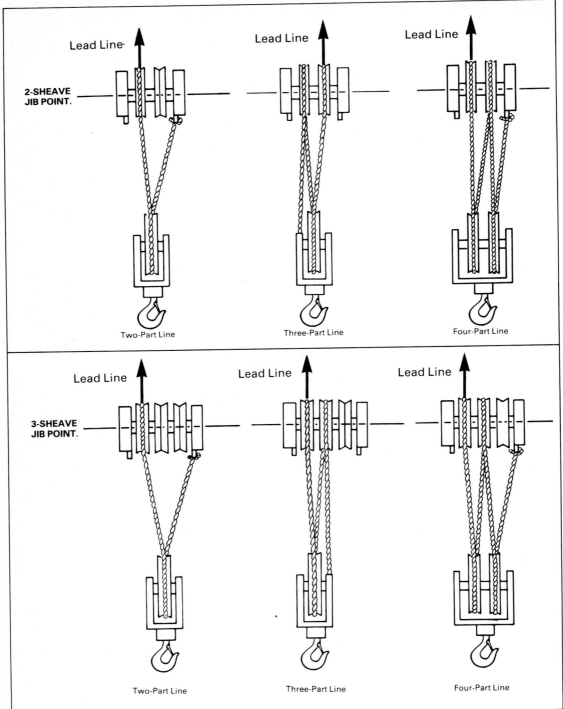

Fig. 5.17 Symmetrical jib point reeving

CHAPTER 6

Slings

Slings require special attention because they are almost always subjected to severe wear, abrasion, impact loading, crushing, kinking and overloading. They also merit special attention because seemingly insignificant changes in sling angle drastically affect the loading. When using slings exercise extreme caution because you are going to be developing unknown loads, in less than ideal circumstances, on less than perfect equipment.

Failure to provide blocking or protective pads will permit sharp corners to cut slings. Pulling slings from under loads will result in abrasion and kinking. Dropping loads on slings or running equipment over them will cause crushing. Sudden starts and stops when lifting loads will increase the stresses in them. Also, improper storage will result in deterioration.

Because of the severe service expected of slings, errors in determining load weight, and the effect of sling angle on the loading, it is recommended that all safe working loads (SWL) be based on a factor of safety of at least 5:1.

SLING CONFIGURATIONS

The term 'sling' includes a wide variety of configurations for all fibre ropes, wire ropes, chains and webs. The most commonly-used types in lifting operations will be considered here because incorrect application can affect the safety of the lift.

Single vertical hitch: (Fig. 6.1) Is a method of supporting a load by a single vertical part or leg of the sling. The total weight of the load is carried by a single leg, the angle of the lift is 90° and the weight of the load can equal the maximum SWL of the sling and fittings. The end fittings of the sling can vary but thimbles should be used in the eyes. Also, the eye splices on wire ropes should be mechanical-Flemish splices for best security. This sling configuration must not be used for lifting loose material, lengthy material or anything that will be difficult to balance. Use them only on items equipped with lifting eye bolts or shackles, such as concrete buckets. They

Fig. 6.1 Single vertical hitch

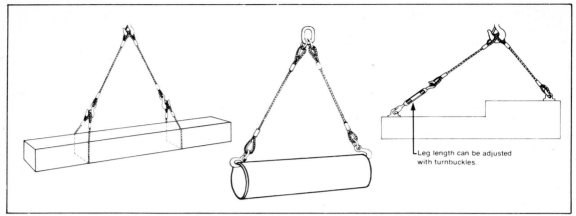

Fig. 6.2 Two-leg bridle hitches

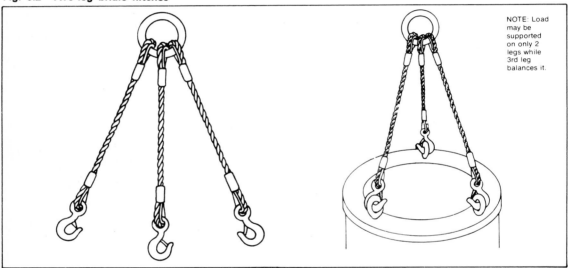

NOTE: Load may be supported on only 2 legs while 3rd leg balances it.

Fig. 6.3 Three-leg bridle hitch

provide absolutely no control over the load because they permit rotation.

Bridle hitch: (Figs. 6.2, 6.3, 6.4) Two, three or four single hitches can be used together to form a bridle hitch for hoisting an object that has the necessary lifting lugs or attachments. The slings can have a wide assortment of end fittings. They provide excellent load stability when the load is distributed equally among the legs, when the hook is directly over the centre of gravity of the load and the load is raised level. In order to distribute the load equally it may be necessary to adjust the leg length with turnbuckles. The use of a bridle sling requires that the sling angles be carefully determined to ensure that the individual legs are not overloaded.

Unless the load is flexible, it is wrong to assume that a three- or four-leg hitch will safely lift a load equal to the safe load on one leg multiplied by the number of legs because there is no way of knowing that each leg is carrying its share of the load. With slings having more than two legs and a rigid load, it is possible for two of the legs to take practically the full load while the others only balance it.

Single basket hitch: (Fig. 6.5) This is a method of supporting a load by placing one end of a sling over a hook, wrapping it around the load and securing the other end to the hook. It cannot be used on any load that is difficult to balance because the load can tilt and slip out of the sling. On loads having inherent stabilising characteristics the load on the sling will be

SLINGS

automatically equalised with each leg supporting half the load. Ensure that the load does not turn or slide along the sling during a lift because both the load and sling will become damaged.

Double basket hitch: (Fig. 6.6) This hitch consists of two single basket hitches passed under the load. They must be placed under the load so that it is properly balanced. The legs of the hitches must be kept far enough apart to provide balance but not so far apart that excessive angles are developed or to create a tendency for the legs to be pulled in toward the centre. On smooth surfaces, both sides of the hitches should be secured against a step or change of contour to prevent the rope from slipping as load is applied. The

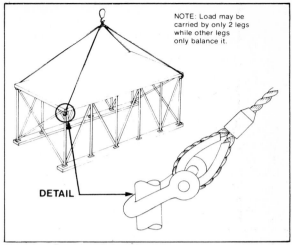

Fig. 6.4 Four-leg bridle hitch

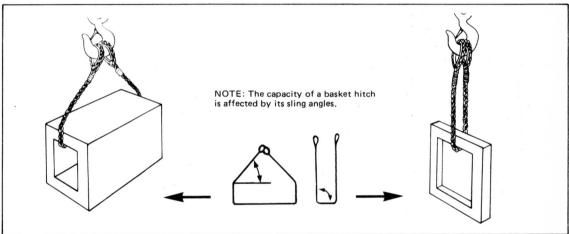

Fig. 6.5 Single basket hitch

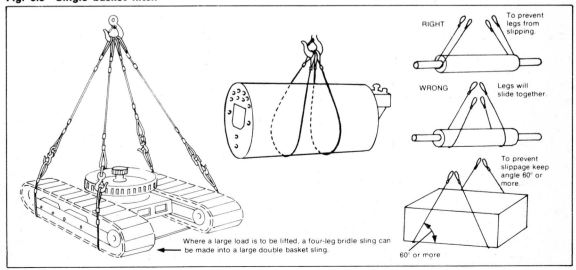

Fig. 6.6 Double basket hitch

angle between the load and the sling should be approximately 60° or greater to avoid slipping.

Double wrap basket hitch: (Fig. 6.7) A basket hitch wrapped completely around the load rather than just supporting it as does the ordinary basket hitch. The double wrap basket hitch can be used in pairs like the double basket hitch. This method is excellent for handling loose material, pipe, rod or smooth cylindrical loads because the rope or chain exerts a full 360° contact with the load and tends to draw it together.

Single choker hitch: (Fig. 6.8) Forms a noose in the rope that tightens as the load is lifted. It does not provide full 360° contact with the load, however, and because of this it should not be used to lift loose bundles from which material can fall or loads that are difficult to balance. The single choker can also be doubled up (not to be confused with double choker hitch), as shown, to provide twice the capacity or to turn a load. When it is necessary to turn a load, the choker is made by placing both eyes of the sling on top of the load with the eyes pointing in the direction opposite to the direction of turn. The centre of the sling is passed around the load, through both eyes and up to the hook. This hitch provides complete control over the load during the entire turning operation, and the load automatically equalises between the two supporting legs of the sling. Because the load is turned into a tight sling, there is no movement between the load and the sling. If it is incorrectly made and the two eyes are placed on the crane hook, the supporting legs of the sling may not be equal in length and the load may be imposed on one leg only.

Double choker hitch: (Fig. 6.9) This consists of two single chokers attached to the load and spread to provide load stability. They, like the single choker, do not completely grip the load but because the load is less likely to tip they are better suited for handling loose bundles, pipes, rods, etc.

Double wrap choker Hitch: (Fig. 6.10) Is one in which the rope or chain is wrapped completely around the load before being hooked into the vertical part of the sling. This hitch is in full contact with the load and tends to draw it tightly together. It can be used either singly on short, easily-balances loads, or in pairs on longer loads.

Endless (or grommet) slings: (Fig. 6.11) These are endless ropes made from one strand of a rope laid or twisted around itself on each successive loop. There is only one tuck in the entire circumference and it is where the two ends enter the rope. These slings can be used in a number of configurations, as vertical hitches, choker hitches and all combinations of these basic configurations. They are very flexible but tend to wear and deteriorate more rapidly than the other slings because normally they are not equipped with fittings and thus are deformed when bent over hooks and bear against themselves at the bight.

Braided slings: (Fig. 6.12) Usually, these are fabricated from six or eight small-diameter ropes braided together to form a single rope that provides a large bearing surface, tremendous strength and flexibility in all directions. They are very easy to handle and almost impossible to kink. The braided sling can be used in all the standard configurations and combinations but is especially useful for basket hitches where low bearing pressure is desirable or where a bend is extremely sharp.

SLING ANGLES

The rated capacity of any sling depends on its size, its configuration and the angles formed by the legs of the sling and the horizontal. A sling with two legs that is used to lift a 450-kg (1000 lb) object will have a 225-kg (500 lb) load in each leg when the sling angle is 90°. The load in each leg will increase as the angle is decreased and at 30° the load will be 450 kg (1000 lb) in each leg. (Fig. 6.13)

If possible, keep the sling angles greater than 45°; sling angles approaching 30° should be considered extremely hazardous and avoided at all costs. The sharp increases in loading at low angles is clearly shown in Fig. 6.14.

Some load tables list sling angles as low as 15° but the use of any sling at an angle less than 30° is extremely dangerous. This is not only because of the high loads associated with them but because of the effect on the load of an error in sling angle measurement of as little as 5°. Table 6.1 illustrates the effect of

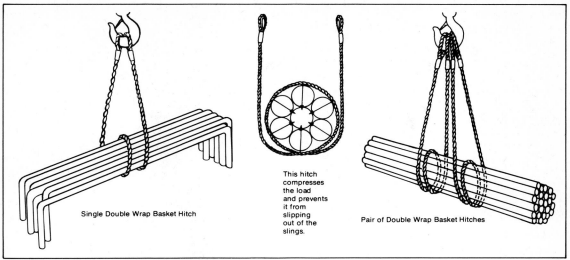

Fig. 6.7 Double wrap basket hitch

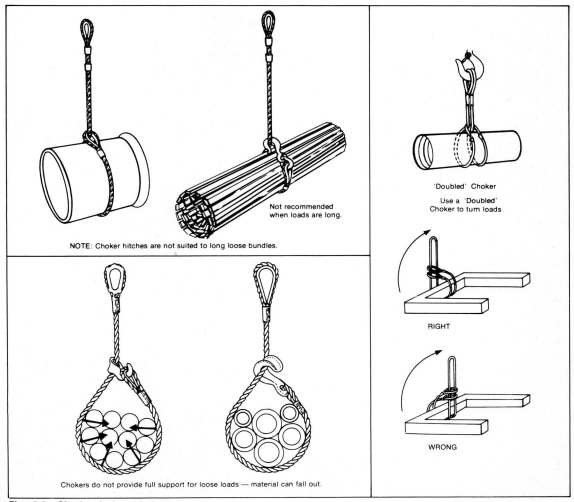

Fig. 6.8 Single choker hitches

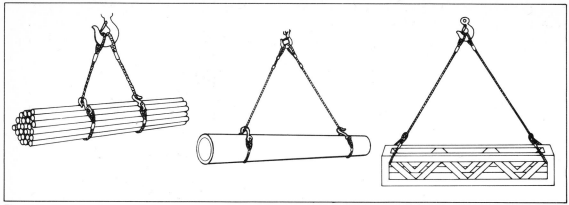

Fig. 6.9 Double choker hitches

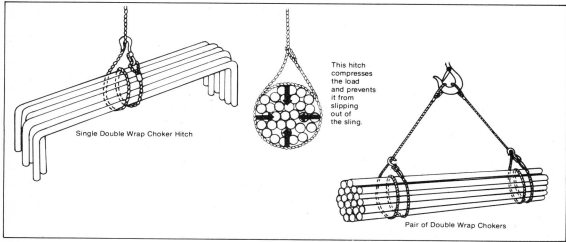

Fig. 6.10 Double wrap choker hitches

TABLE 6.1

EFFECT ON LOADS OF SLING ANGLE MEASUREMENT ERROR					
Assumed sling angle	Assumed load per leg		Actual angle (is 5° less than assumed angle)	Actual load per leg		Error %
	kg	lb		kg	lb	
90°	226	500	85°	227	502	0.4
75°	235	518	70°	241	532	2.8
60°	262	577	55°	276	610	5.7
45°	320	707	40°	352	778	9.1
30°	453	1000	25°	537	1183	18.3
15°	875	1932	10°	1305	2880	49.0

a 5° error in sling-angle measurement on the sling load.

You can see that there is almost a 50% error in the assumed load at the 15° sling angle. This illustrates how important it is not only to ensure that the sling angle is greater than 45° but also to measure it accurately. The easiest and most accurate way to determine the angle is by measuring it with a large plywood measure graduated in degrees.

SAFE WORKING LOADS (SWL)

The remaining sections of this chapter contain many tables of SWL. It would be a very difficult task to remember all load, size and sling-angle combinations, but the following rules-of-thumb work well for estimating the loads in the most common sling configurations.

Each of the rules-of-thumb for a given sling configuration, material and size, is based on the SWL of the single vertical hitch of that sling. The efficiencies of whatever end fittings

SLINGS

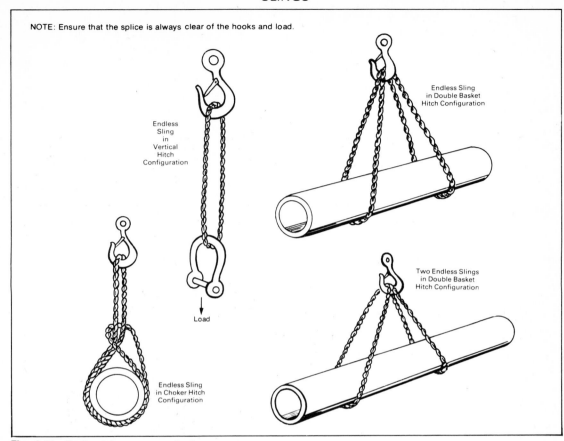

Fig. 6.11 Endless (or grommet) slings

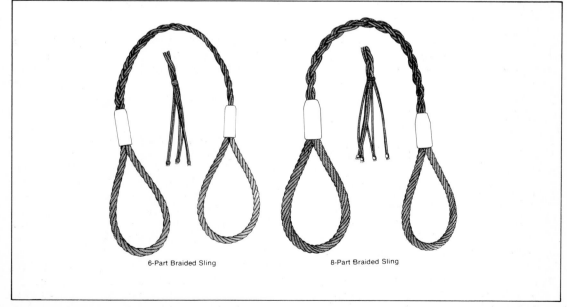

Fig. 6.12 Braided slings

138 SLINGS

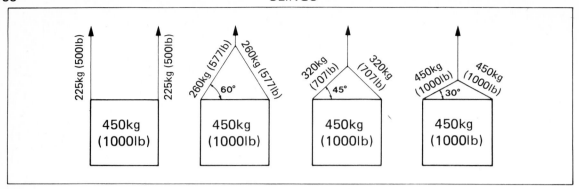

Fig. 6.13 How sling angle affects sling load

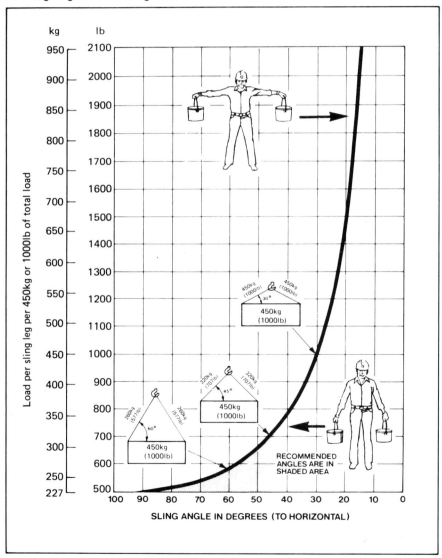

Fig. 6.14 Graph showing how sling angle affects sling load

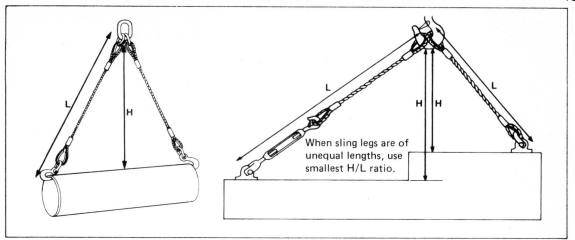

Fig. 6.15 Determining capacity of two-leg bridle hitch

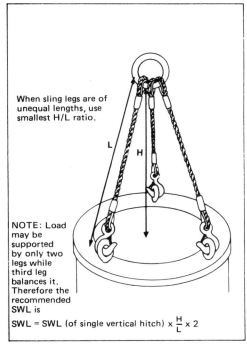

Fig. 6.16 Determining capacity of three-leg bridle hitch

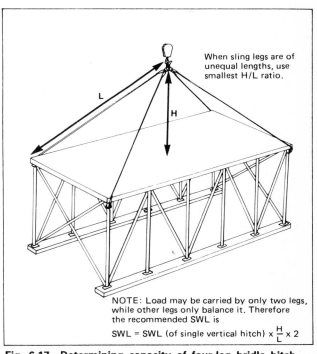

Fig. 6.17 Determining capacity of four-leg bridle hitch

are used will also have to be taken into account when determining the capacity of the combination.

– *Bridle hitches* (two-, three- and four-leg): (Figs. 6.15, 6.16, 6.17)
Measure the length of the sling legs (=L) and measure the headroom between the hook and the load (=H).

$$\text{SWL} = \text{SWL (of single vertical hitch)} \times \frac{H}{L} \times 2.$$

This formula is for a two-leg bridle hitch but it is strongly recommended that it also be used for the three- and four-leg hitches. It is wrong to assume that a three- or four-leg hitch will safely lift a load equal to the safe load on one leg multiplied by the number of legs because there is no way of knowing that each leg is carrying its share of the load. With slings having more than two legs for a rigid load, it is possible for two of the legs to take practically the full load while the others only balance it.

– *Single basket hitch:* (Fig. 6.18)

> For vertical legs – SWL = SWL (of single vertical hitch) × 2.
>
> For inclined legs – SWL = SWL (of single vertical hitch) × $\frac{H}{L}$ × 2.

– *Double basket hitch:* (Fig. 6.19)

> For vertical legs – SWL = SWL (of single vertical hitch) × 4.
>
> For inclined legs – SWL = SWL (of single vertical hitch) × $\frac{H}{L}$ × 4.

– *Double wrap basket hitch:*
Depending on the configurations, the SWL are the same as for the single basket hitch or the double basket hitch.

– *Single choker hitch:* (Fig. 6.20)

> For sling angles of 45° or more:
>
> SWL = SWL (of single vertical hitch) × ¾.
>
> Sling angles of less than 45° are not recommended. However, if they must be used the formula is:
>
> SWL = SWL (of single vertical hitch) × A/B.

– *Double choker hitch:* (Fig. 6.21)

> For sling angles of 45° or more (formed by the choker),
>
> SWL = SWL (of single vertical hitch) × $\frac{3}{4} \times \frac{H}{L} \times 2$
>
> Sling angles of less than 45° (formed by the choker) are not recommended. However, if they must be used the formula is:
>
> SWL = SWL (of single vertical hitch) × $\frac{A}{B} \times \frac{H}{L} \times 2$.

– *Double wrap choker hitch:*
Depending on the configuration, the SWL are the same as for the single choker hitch or the double choker hitch.

– *Endless (or grommet) slings:*
Depending on the configuration, the SWL are twice the values for the previous configurations.

FIBRE ROPE SLINGS

Fibre rope slings are preferred for some applications because they are pliant, they grip the load well and they do not mar the surface of the load. They should be used only on light loads, however, and must not be used on

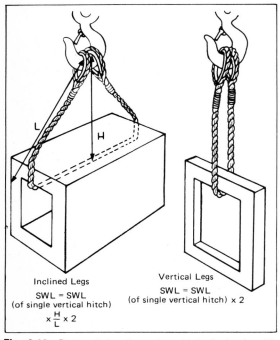

Fig. 6.18 Determining capacity of single basket hitch

Inclined Legs
SWL = SWL (of single vertical hitch) × $\frac{H}{L}$ × 2

Vertical Legs
SWL = SWL (of single vertical hitch) × 2

SLINGS

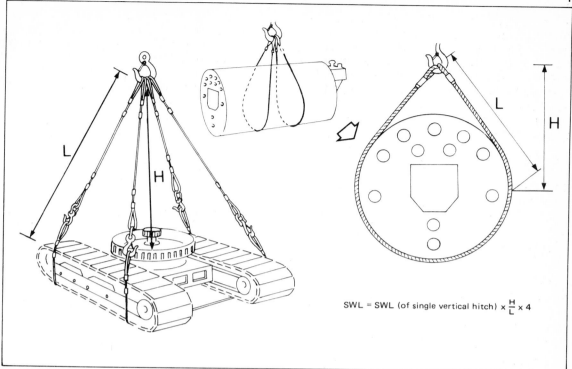

Fig. 6.19 Determining capacity of double basket hitch with inclined legs

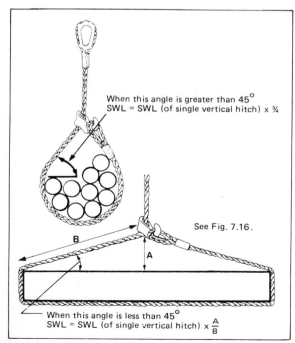

Fig. 6.20 Determining capacity of single choker hitch

objects that have sharp edges capable of cutting the rope or in applications where the sling will be exposed to high temperatures, severe abrasion or acids.

The choice of the rope type and size will depend upon the application, the weight to be lifted and the sling angle. Before lifting any load with a fibre rope sling be sure to inspect the sling carefully because this type deteriorates far more rapidly than wire rope slings and the actual strength is very difficult to estimate.

The SWL (Canadian standard) in Tables 6.2, 6.3, 6.4 and 6.5 are approximate only and are based on a 5:1 factor of safety. They refer to new unknotted ropes and they assume that the loads are symmetrical and balanced, that the sling legs are of equal length, that the rope splices and fittings used to couple the slings to the load are strong enough and that padding is used to protect the slings when lifting objects with sharp corners.

SLINGS
TABLE 6.2
(Canadian Standard)

MANILA ROPE SLINGS
Spliced eyes in both ends

MAXIMUM SAFE WORKING LOADS (SWL) – tonnef
(Safety Factor = 5)

Rope diameter		Single vertical hitch	Single choker hitch	Single basket hitch (vertical legs)	Two-leg bridle hitch and single basket hitch with legs inclined		
mm	in				60°	45°	30°
10	3/8	0.12	0.09	0.24	0.21	0.17	0.12
13	1/2	0.24	0.18	0.48	0.42	0.34	0.24
14	9/16	0.31	0.23	0.62	0.54	0.44	0.31
16	5/8	0.40	0.30	0.80	0.69	0.56	0.40
19	3/4	0.49	0.36	0.98	0.85	0.69	0.49
21	13/16	0.59	0.44	1.18	1.02	0.83	0.59
22	7/8	0.70	0.52	1.40	1.21	0.99	0.70
25	1	0.82	0.61	1.63	1.41	1.16	0.82
27	1 1/16	0.95	0.72	1.90	1.63	1.35	0.95
29	1 1/8	1.09	0.82	2.18	1.88	1.54	1.09
32	1 1/4	1.22	0.91	2.45	2.12	1.72	1.22
33	1 5/16	1.36	1.02	2.72	2.36	1.90	1.36
38	1 1/2	1.68	1.27	3.36	2.90	2.38	1.68
41	1 5/8	2.04	1.54	4.08	3.54	2.88	2.04
44	1 3/4	2.40	1.81	4.81	4.17	3.40	2.40
51	2	2.81	2.11	5.62	4.85	3.99	2.81
54	2 1/8	3.27	2.45	6.53	5.67	4.63	3.27
57	2 1/4	3.72	2.79	7.44	6.44	5.26	3.72
64	2 1/2	4.22	3.17	8.44	7.30	5.99	4.22
67	2 5/8	4.72	3.53	9.43	8.16	6.67	4.72

If used with Double Choker Hitch multiply above values by 3/4

For Double Basket Hitch multiply above values by 2.

Note: For safe working loads of endless or grommet slings, multiply above values by 2.
See also BS.2052 (Natural Fibre Ropes).

SLINGS
TABLE 6.3
(Canadian Standard)

NYLON ROPE SLINGS
Spliced eyes in both ends

MAXIMUM SAFE WORKING LOADS (SWL) – tonnef
(Safety Factor = 5)

Rope diameter		Single vertical hitch	Single choker hitch	Single basket hitch (vertical legs)	Two-leg bridle hitch and single basket hitch with legs inclined		
mm	in				60°	45°	30°
10	3/8	0.32	0.24	0.63	0.54	0.45	0.32
13	1/2	0.54	0.43	1.13	1.00	0.80	0.54
14	9/16	0.68	0.51	1.36	1.18	0.95	0.68
16	5/8	0.91	0.68	1.81	1.59	1.27	0.91
19	3/4	1.27	0.95	2.54	2.20	1.81	1.27
21	13/16	1.45	1.09	2.99	2.49	2.04	1.45
22	7/8	1.72	1.29	3.45	2.99	2.45	1.72
25	1	2.18	1.63	4.35	3.76	3.08	2.18
27	1 1/16	2.49	1.87	4.99	4.31	3.53	2.49
29	1 1/8	2.86	2.14	5.71	4.94	4.04	2.86
32	1 1/4	3.27	2.45	6.53	5.67	4.63	3.27
33	1 5/16	3.72	2.79	7.44	6.44	5.26	3.72
38	1 1/2	4.63	3.47	9.25	8.03	6.53	4.63
41	1 5/8	5.62	4.22	11.25	9.75	7.94	5.62
44	1 3/4	6.80	5.10	13.60	11.79	9.61	6.80
51	2	8.12	6.08	16.24	14.06	11.47	8.12
54	2 1/8	9.16	6.37	18.32	15.87	12.97	9.16
57	2 1/4	10.79	8.09	21.50	18.68	15.28	10.79
64	2 1/2	12.06	9.07	24.13	20.91	17.05	12.06
67	2 5/8	13.92	10.43	27.84	24.13	19.68	13.92

If used with Double Choker Hitch multiply above values by 3/4

For Double Basket Hitch multiply above values by 2.

Note: For safe working loads of endless or grommet slings, multiply above values by 2.
See also BS.4928 (Man-made Fibre Ropes).

SLINGS
TABLE 6.4
(Canadian Standard)

POLYPROPYLENE ROPE SLINGS
Spliced eyes in both ends

MAXIMUM SAFE WORKING LOADS (SWL) – tonnef
(Safety Factor = 5)

Rope diameter		Single vertical hitch	Single choker hitch	Single basket hitch (vertical legs)	Two-leg bridle hitch and single basket hitch with legs inclined		
mm	in				60°	45°	30°
10	3/8	0.23	0.17	0.45	0.39	0.32	0.23
13	1/2	0.38	0.28	0.75	0.63	0.54	0.38
14	9/16	0.44	0.33	0.87	0.77	0.61	0.44
16	5/8	0.59	0.44	1.18	1.02	0.82	0.59
19	3/4	0.77	0.58	1.54	1.32	1.09	0.77
21	13/16	0.86	0.66	1.72	1.50	1.22	0.86
22	7/8	1.00	0.75	2.00	1.72	1.41	1.00
25	1	1.32	0.99	2.63	2.27	1.86	1.32
27	1 1/16	1.36	1.02	2.72	2.36	1.90	1.36
29	1 1/8	1.70	1.27	3.40	2.95	2.40	1.70
32	1 1/4	1.90	1.43	3.81	3.31	2.68	1.90
33	1 5/16	2.00	1.50	3.99	3.45	2.81	2.00
38	1 1/2	2.72	2.04	5.44	4.72	3.85	2.72
41	1 5/8	3.31	2.49	6.62	5.71	4.67	3.31
44	1 3/4	3.95	2.95	7.89	6.85	5.58	3.95
51	2	4.72	3.54	9.43	8.16	6.67	4.72
54	2 1/8	5.22	3.90	10.43	9.02	7.39	5.22
57	2 1/4	5.99	4.49	11.97	10.39	8.48	5.99
64	2 1/2	6.85	5.12	13.70	11.88	9.70	6.85
67	2 5/8	7.71	5.78	15.42	13.33	10.88	7.71

If used with Double Choker Hitch multiply above values by 3/4

For Double Basket Hitch multiply above values by 2.

Note: For safe working loads of endless or grommet slings, multiply above values by 2.
See also BS.4928 (Man-made Fibre Ropes).

SLINGS
TABLE 6.5
(Canadian Standard)

POLYESTER ROPE SLINGS
Spliced eyes in both ends

MAXIMUM SAFE WORKING LOADS (SWL) – tonnef
(Safety Factor = 5)

Rope diameter		Single vertical hitch	Single choker hitch	Single basket hitch (vertical legs)	Two-leg bridle hitch and single basket hitch with legs inclined		
mm	in				60°	45°	30°
10	3/8	0.32	0.24	0.63	0.54	0.45	0.32
13	1/2	0.54	0.41	1.09	0.95	0.77	0.54
14	9/16	0.68	0.51	1.36	1.18	0.95	0.68
16	5/8	0.86	0.65	1.72	1.50	1.22	0.86
19	3/4	1.09	0.82	2.18	1.88	1.54	1.09
21	13/16	1.34	1.00	2.68	2.31	1.90	1.34
22	7/8	1.54	1.16	3.08	2.68	2.18	1.54
25	1	1.90	1.43	3.81	3.31	2.68	1.90
27	1 1/16	2.22	1.67	4.44	3.85	3.13	2.22
29	1 1/8	2.54	1.90	5.08	4.40	3.58	2.54
32	1 1/4	2.86	2.14	5.71	4.94	4.04	2.86
33	1 5/16	3.22	2.41	6.44	5.58	4.53	3.22
38	1 1/2	4.04	3.03	8.07	6.98	5.71	4.04
41	1 5/8	4.90	3.67	9.80	8.48	6.94	4.90
44	1 3/4	5.85	4.39	11.70	10.11	8.25	5.85
51	2	6.89	5.17	13.79	11.93	9.75	6.89
54	2 1/8	7.89	5.92	15.78	13.65	11.16	7.89
57	2 1/4	9.25	6.94	18.50	16.00	13.06	9.25
64	2 1/2	10.52	7.80	21.04	18.23	14.87	10.52
67	2 5/8	11.79	8.84	23.58	20.40	16.69	11.79

If used with Double Choker Hitch multiply above values by 3/4

For Double Basket Hitch multiply above values by 2.

Note: For safe working loads of endless or grommet slings, multiply above values by 2.
See also BS.4928 (Man-made Fibre Ropes).

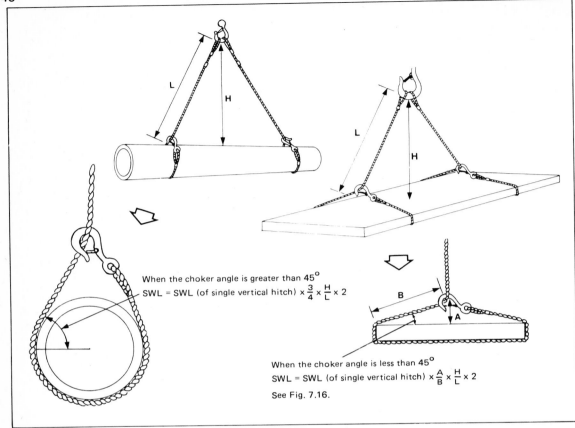

Fig. 6.21 Determining capacity of double choker hitch

SYNTHETIC WEBBING SLINGS

Synthetic webbing slings offer a number of advantages for lifting purposes:

— Their width and relative softness mean they have much less tendency to mar or scratch finely machined, highly polished or painted surfaces and have less tendency to crush fragile objects than do fibre rope, wire rope or chain slings. (Fig. 6.22)
— Because of their flexibility, they tend to mould themselves to the shape of the load. (Fig. 6.23)
— They are not affected by moisture and certain chemicals.
— They do not rust and thus do not stain ornamental precast concrete or stone.
— They are non-sparking and can be used safely in explosive atmospheres.
— They minimise twisting and spinning during lifting.
— Their light weight permits ease of fitting, their softness precludes hand cuts, and the danger of harm from a bump by a free-swinging sling is minimal.
— They are elastic and stretch under load more than either wire rope or chain and are thus able to absorb heavy shocks and cushion the load. In cases where sling stretching must be minimised, a sling of larger load capacity or a polyester sling should be used.

SLINGS

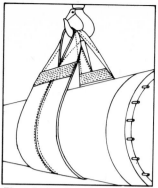

Fig. 6.22 Synthetic web slings do not damage or crush the load as do wire ropes or chain

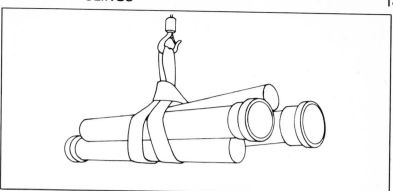

Fig. 6.23 Pipe handling shows how webbing slings can mould themselves to the load, thus allowing secure handling of irregularly-shaped loads

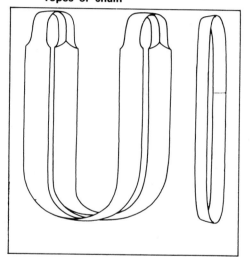

Fig. 6.24 Endless (or grommet) sling

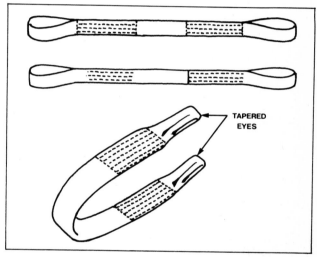

Fig. 6.25 Standard eye-and-eye slings

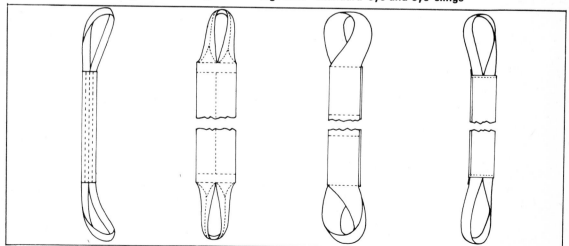

Fig. 6.26 Twisted-eye slings

Synthetic webbing slings are available in a number of configurations that find application in the industry.

- *Endless (or grommet):* (Fig. 6.24) Both ends of one piece of webbing are lapped and sewn to form a continuous piece. They can be used as vertical hitches, bridle hitches, in choker arrangements or as basket slings. Because load contact points can be shifted with every lift, wear is evenly distributed and sling life is extended.
- *Standard eye-and-eye:* (Fig. 6.25) Webbing asembled and sewn to form a flat body sling with an eye at each end and the eye openings in the same plane as the sling body. The eyes may either be full web width or may be tapered by being folded and sewn to a width narrower than the webbing width.
- *Twisted-eye:* (Fig. 6.26) An eye-and-eye type with twisted terminations at both ends. The eye openings are at 90° to the plane of the sling body. This configuration is also available with either full width or tapered eyes.

Slings are also available with metal end fittings in place of the sewn eyes, the most common being triangle and choker equipment. Combination equipment consists of a triangle for one end of the sling and a triangle/rectangle (choker attachment) for the other end. With this arrangement, both choker and basket as well as straight hitches may be rigged. They help reduce wear in the sling eyes and thus lengthen sling life. (Fig. 6.27)

Despite their inherent toughness, synthetic webbing slings can be cut by repeated use around sharp-cornered objects and they eventually show signs of abrasion when rough-surfaced products are continually hoisted. There are, however, protective devices offered by most sling manufacturers that minimise these effects. (Fig. 6.28)

- Buffer strips of leather, nylon, or other materials sewn on the body of a sling protect against wear. Leather pads are the most resistant to wear and cutting, but are subject to weathering and gradual deterioration. They are not recommended in lengths over 2 m (6 ft) due to the different stretch characteristics of leather and webbing. On the other hand, nylon web wear pads are more resistant to weathering, oils, grease, most alkalies; and they stretch in the same ratio as the sling body.
- Edge guards are also available and consist of strips of webbing or leather sewn around each edge of the sling. This is necessary for certain applications where the sling edges are subject to damage.
- Sleeve or sliding-tube type wear pads are available for slings used to handle material having sharp edges. They can be positioned on the sling where required, don't move when the sling stretches, adjust to the load and cover both sides of the sling.
- Reinforcing strips can be sewn into the sling eyes that double or triple the eye thickness and greatly increase the life and safety.
- Coatings can be applied to provide added abrasion and chemical resistance. These treatments also increase the coefficient of friction, affording a better grip when loads with slippery surfaces are to be handled. These coatings can be brightly coloured for safety or load rating purposes.
- Some manufacturers offer a cotton-faced nylon webbing for hoisting granite and other rough surfaced material.

The sling capacities quoted in Tables 6.6, 6.7 and 6.8 of Canadian Standards are approximate only and are based on nylon webbing having breaking strengths of 1400 kN/m (8000 lb/in) of webbing width and 1050 kN/m (6000 lb/in) of webbing width and Dacron having a breaking strength of 875 kN/m (5000 lb/in) of webbing width. The capacities are also based on a 5:1 factor of safety and assume that the end fittings are of adequate strength.

Although safe working loads for bridle hitches in the choker or double basket configuration hitches are shown, it is advised that they be used only with extreme caution because as the sling angle decreases one edge of the web will take all the load with risk of tearing. (Fig. 6.29)

The importance of inspection of the slings must not be underestimated even though in general, if they look good they probably are good. Any damage is usually easy to detect; worn eyes and worn or distorted fittings, cuts, holes, punches, tears, frayed material, broken stitching and acid, caustic or heat burns – all are immediately evident and indicate that the slings should be replaced. Do not attempt to repair them yourself.

SLINGS

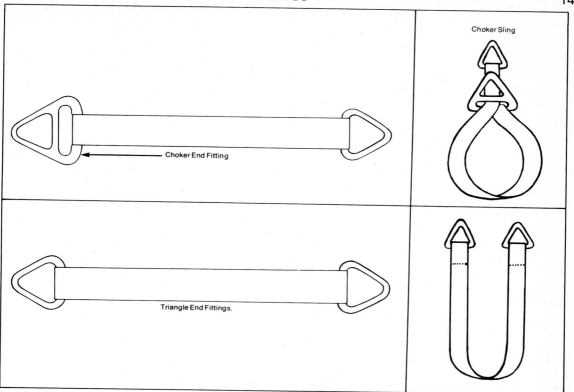

Fig. 6.27 Metal end fittings

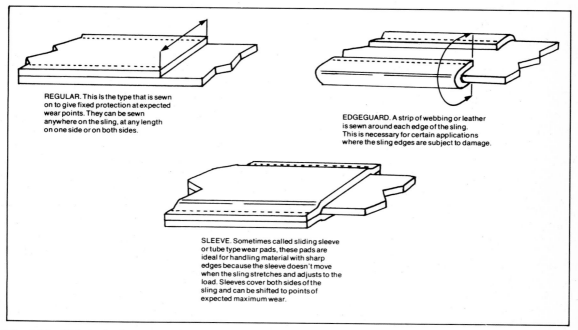

Fig. 6.28 Web and edge protectors

TABLE 6.6
(Canadian Standard)

NYLON WEB SLINGS
Breaking Force 1400 kN/m (8000 lb per in width) material

MAXIMUM SAFE WORKING LOADS (SWL) – tonnef (Safety Factor = 5)
(Eye-and-eye, twisted eye, triangle fittings, choker fittings)

Web width		Single vertical hitch	Single choker hitch	Single basket hitch (vertical legs)	Two-leg bridle hitch and single basket hitch with legs inclined		
mm	in				60°	45°	30°
25	1	0.73	0.54	1.45	1.26	1.03	0.73
51	2	1.45	1.09	2.90	2.52	2.05	1.45
76	3	2.18	1.63	4.35	3.76	3.08	2.18
102	4	2.90	2.18	5.81	5.03	4.11	2.90
127	5	3.63	2.72	7.26	6.28	5.13	3.63
152	6	4.35	3.27	8.71	7.53	6.17	4.35
178	7	5.08	3.81	10.16	8.80	7.17	5.08
204	8	5.81	4.35	11.61	10.07	8.21	5.81
229	9	6.53	4.90	13.06	11.34	9.25	6.53
254	10	7.25	5.44	14.51	12.56	10.25	7.25
279	11	7.98	5.99	15.97	13.83	11.29	7.28
305	12	8.71	6.53	17.42	15.10	12.34	8.71

If used with Double Choker Hitch multiply above values by ¾

For Double Basket Hitch multiply above values by 2.

Note: For safe working loads of endless or grommet slings, multiply above values by 2.
See also BS.3481 Part 2 (Flat Lifting Slings – Man-made Fibre): Specific breaking force in kN per 100 mm of width

SLINGS

TABLE 6.7
(Canadian Standard)

NYLON WEB SLINGS
Breaking Force 1050 kN/m (6000lb per in width) material

MAXIMUM SAFE WORKING LOADS (SWL) – tonnef (Safety Factor = 5)
(Eye-and-eye, twisted eye, triangle fittings, choker fittings)

Web width		Single vertical hitch	Single choker hitch	Single basket hitch (vertical legs)	Two-leg bridle hitch and single basket hitch with legs inclined		
mm	in				60°	45°	30°
25	1	0.54	0.41	1.09	0.94	0.77	0.54
51	2	1.09	0.82	2.18	1.89	1.54	1.09
76	3	1.63	1.22	3.27	2.83	2.31	1.63
102	4	2.18	1.63	4.35	3.76	3.08	2.18
127	5	2.72	2.04	5.44	4.72	3.85	2.72
152	6	3.27	2.45	6.53	5.67	4.63	3.27
178	7	3.81	2.86	7.62	6.60	5.40	3.81
204	8	4.35	3.27	8.71	7.53	6.17	4.35
229	9	4.90	3.67	9.80	8.48	6.94	4.90
254	10	5.44	4.08	10.88	9.43	7.71	5.44
279	11	5.99	4.49	11.97	10.39	8.46	5.99
305	12	6.53	4.90	13.06	11.34	9.25	6.53

If used with Double Choker Hitch multiply above values by ¾

For Double Basket Hitch multiply above values by 2.

Note: For safe working loads of endless or grommet slings, multiply above values by 2.
See also BS.3481 Part 2 (Flat Lifting Slings – Man-made Fibre): Specific breaking force in kN per 100 mm of width

TABLE 6.8
(Canadian Standard)

DACRON WEB SLINGS
Breaking Force 875 kN/m (5000 lb per in width) material

MAXIMUM SAFE WORKING LOADS (SWL) – tonnef (Safety Factor = 5)
(Eye-and-eye, twisted eye, triangle fittings, choker fittings)

Web width		Single vertical hitch	Single choker hitch	Single basket hitch (vertical legs)	Two-leg bridle hitch and single basket hitch with legs inclined		
mm	in				60°	45°	30°
25	1	0.45	0.34	0.90	0.78	0.63	0.45
51	2	0.90	0.68	1.81	1.57	1.28	0.90
76	3	1.36	1.02	2.72	2.36	1.93	1.36
102	4	1.81	1.36	3.63	3.15	2.56	1.82
127	5	2.27	1.70	4.53	3.93	3.21	2.26
152	6	2.72	2.04	5.44	4.72	3.85	2.72
178	7	3.17	2.38	6.35	5.49	4.49	3.18
204	8	3.63	2.72	7.26	6.28	5.12	3.63
229	9	4.08	3.06	8.16	7.07	5.76	4.08
254	10	4.53	3.40	9.07	7.87	6.39	4.54
279	11	4.99	3.74	9.98	8.66	7.03	4.99
305	12	5.44	4.08	10.88	9.43	7.71	5.44

If used with Double Choker Hitch multiply above values by ¾

For Double Basket Hitch multiply above values by 2.

Note: For safe working loads of endless or grommet slings, multiply above values by 2.
See also BS.3481 Part 2 (Flat Lifting Slings – Man-made fibre): Specific breaking force in kN per 100 mm of width

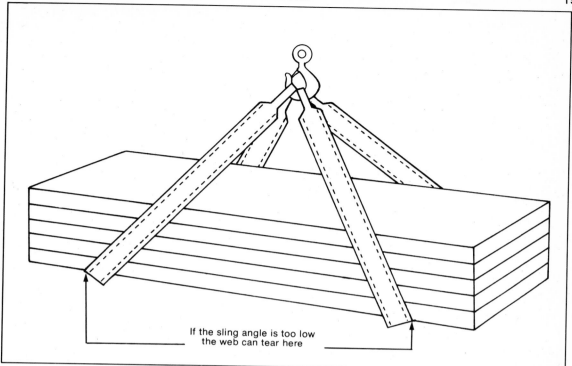

Fig. 6.29 Effect on webbing of low sling angle

METAL (WIRE OR CHAIN) MESH SLINGS

Wire or metal mesh slings are well adapted for use where the loads are abrasive or hot, or would tend to cut fabric slings and wire ropes. They resist abrasion and cutting, grip the load firmly without stretching and can withstand temperatures of up to 290°C (550°F). They have smooth, flat bearing surfaces, they conform to irregular shapes, do not kink or tangle and resist corrosion. (Fig. 6.30)

They are available in the following three mesh sizes:

- 10 gauge – (heavy duty) recommended for general purpose lifting because it combines strength and abrasion-resistance with flexibility.
- 12 gauge – for medium duty applications.
- 14 gauge – for very light duty.

For handling materials that would damage the wire mesh, or for handling loads with finishes that the mesh would damage, the slings can be coated with rubber or plastic. See Table 6.9 of Canadian standards for their SWL.

CHAIN SLINGS

Chain slings find application in those areas where the primary requirements are ruggedness, abrasion resistance and high temperature resistance.

The only chain suitable for overhead lifting is fabricated from alloy steel and identified as shown in Fig. 3.1. See Chapter 3 for guidelines on the safe use of chain.

They should be padded where they bear on sharp edges of metal parts otherwise some of the links will be subjected to bending stresses for which they were not designed. The chain may also damage the material being lifted if it is not padded.

Whenever a chain sling is hooked back on itself (into the chain rather than the master

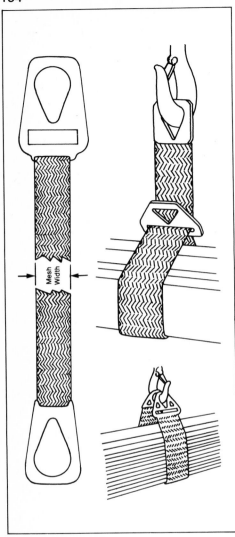

Fig. 6.30 Chain mesh slings

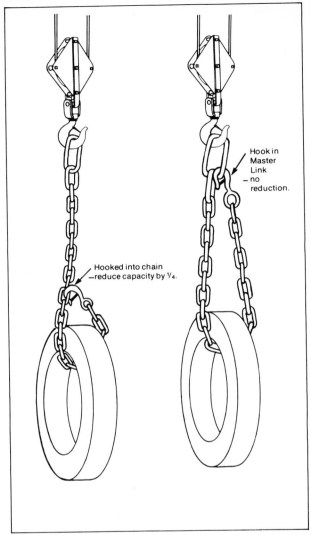

Fig. 6.31 Chain slings

link or master coupling link) the capacity of the sling should be reduced by ¼. For this reason, the choker hitch ratings in Table 6.10 are ¼ less than the vertical hitch ratings. Whenever using chain slings in the basket hitch configuration, ensure that they are hooked back into the master link, not the chain or the coupling link, and also ensure that the master link is capable of carrying the imposed load (Fig. 6.31). Table 6.10 gives their rated SWL to BS.3458.

WIRE ROPE SLINGS

The use of steel wire rope slings for lifting materials provides several advantages over other materials. Great strength and a minimum of weight are combined with flexibility. Before failure warning is given by the breaking of outer wires. Reserve strength is provided in that the inner wires are protected by the outer wires and possess sufficient strength to carry the load if a reasonable factor of safety is allowed for the sling.

Correctly fabricated wire rope slings are the safest types available for general construction usage. They do not wear as rapidly as fibre rope slings. Wire rope slings show, by inspection, their true condition and the presence of broken wires clearly indicates the extent of fatigue, wear, abrasion and the like.

However, they also have limitations and must be carefully selected for the application and service in which they are to be placed. The procedures and precautions outlined in the chapter on wire rope must be followed regardless of the sling type or application.

All wire rope slings should be made of improved plow steel and should have independent wire rope cores to reduce the possibility of the rope being crushed in service.

It is recommended that all eyes in wire rope slings be equipped with thimbles and that the eyes be formed with the Flemish splice and secured by swaged or pressed mechanical sleeves or fittings. With the exception of socketed connections, this is the *only* method that produces an eye as strong as the rope itself and provides reserve strength should the mechanical sleeve or fitting fail or loosen.

Tables 6.11–6.21 are provided for information only on the more commonly-used types of wire rope slings based on Canadian standards. For actual safe working loads of slings consult with the sling manufacturer. These tables are based on safety factors of 5:1 and assume that the slings are in excellent condition. Any deterioration, such as described in the chapter on wire rope, reduces the capacity of the sling.

SLINGS

TABLE 6.9
(Canadian Standard)

METAL (WIRE OR CHAIN) MESH SLINGS

MAXIMUM SAFE WORKING LOADS (SWL) – tonnef
(Safety Factor = 5)

Sling width		Single vertical hitch	Single choker hitch	Single basket hitch (vertical legs)	Two-leg bridle hitch and single basket hitch with legs inclined		
mm	in				60°	45°	30°
HEAVY DUTY CLASSIFICATION (10 GAUGE MESH)							
51	2	0.68	0.50	1.36	1.18	0.95	0.68
76	3	1.23	0.91	2.45	2.13	1.72	1.23
102	4	1.81	1.36	3.63	3.13	2.54	1.81
152	6	2.72	2.04	5.44	4.72	3.86	2.72
203	8	3.63	2.72	7.26	6.26	5.13	3.63
254	10	4.54	3.40	9.07	7.85	6.40	4.54
305	12	5.44	4.08	10.89	9.44	7.71	5.44
MEDIUM DUTY CLASSIFICATION (12 GAUGE MESH)							
51	2	0.61	0.45	1.23	1.04	0.86	0.61
76	3	0.91	0.68	1.81	1.59	1.27	0.91
102	4	1.23	0.91	2.45	2.13	1.72	1.23
152	6	2.04	1.54	4.08	3.54	2.90	2.04
203	8	2.72	2.04	5.44	4.72	3.86	2.72
254	10	3.40	2.54	6.80	5.90	4.81	3.40
305	12	4.08	3.06	8.17	7.08	5.76	4.08
LIGHT DUTY CLASSIFICATION (14 GAUGE)							
51	2	0.41	0.32	0.82	0.73	0.59	0.41
76	3	0.64	0.45	1.27	1.09	0.91	0.64
102	4	0.91	0.68	1.81	1.59	1.27	0.91
152	6	1.36	1.02	2.72	2.36	1.91	1.36
203	8	1.81	1.36	3.63	3.13	2.59	1.81
254	10	2.27	1.70	4.54	3.90	3.22	2.27
305	12	2.72	2.04	5.44	4.72	3.86	2.72

If used with Double Choker Hitch multiply above values by ¾

For Double Basket Hitch multiply above values by 2.

See also BS.3481 Part 1 (Wire Coil Flat Slings).

SLINGS

TABLE 6.10

Chain size		Single vertical leg	MAXIMUM SAFE WORKING LOADS – tonnef				
mm	in		0°	30°	60°	90°	120°
6	¼	0.76	1.52	1.47	1.32	1.07	0.76
10	⅜	1.68	3.36	3.20	2.90	2.34	1.68
13	½	3.05	6.10	5.89	5.28	4.27	3.05
16	⅝	4.72	9.45	9.09	8.18	6.65	4.72
19	¾	6.86	13.72	13.21	11.84	9.70	6.86
22	⅞	9.30	18.60	17.93	16.10	13.11	9.30
25	1	12.19	24.38	23.52	21.08	17.22	12.19
29	1⅛	15.39	30.78	29.72	26.62	21.74	15.39
32	1¼	19.05	38.10	36.83	32.97	26.92	19.05
35	1⅜	23.01	46.02	44.45	39.83	32.50	23.01
38	1½	27.43	54.86	52.96	46.89	38.76	27.43
Loading factor (90° angle between legs)			1.41	1.36	1.22	1.00	0.70
% increase or reduction			+40	+35	+20	—	−30

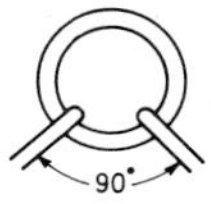

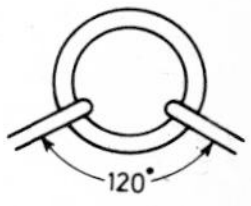

CHAIN SLINGS (ALLOY STEEL TO BS.3458)

For Choker Hitch multiply SWL by ¾.
For Double Basket Hitch multiply SWL by 2.

SLINGS

TABLE 6.11
(Canadian Standard)

WIRE ROPE SLINGS
6 × 19 Classification Group, Improved Plow Steel, Fibre Core

MAXIMUM SAFE WORKING LOADS (SWL) – tonnef
(Safety Factor = 5)

Rope diameter		Single vertical hitch 	Single choker hitch 	Single basket hitch (vertical legs) 	Two-leg bridle hitch and single basket hitch with legs inclined 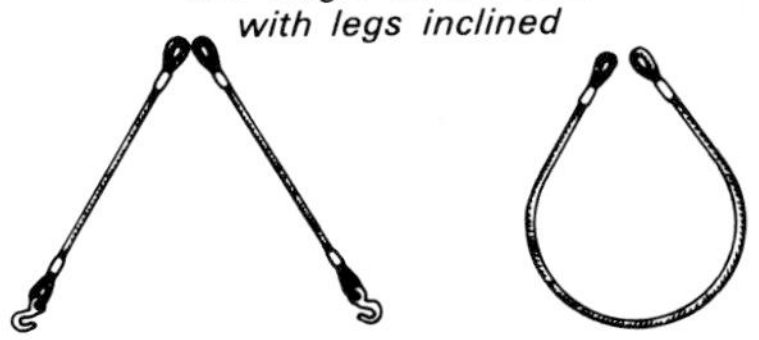		
mm	in				60°	45°	30°
5	3/16	0.27	0.20	0.54	0.48	0.39	0.27
6	1/4	0.50	0.37	1.00	0.86	0.70	0.50
8	5/16	0.75	0.57	1.50	1.29	1.07	0.75
9	3/8	1.09	0.82	2.18	1.88	1.54	1.09
11	7/16	1.45	1.09	2.90	2.52	2.04	1.45
13	1/2	2.00	1.50	3.99	3.45	2.81	2.00
14	9/16	2.40	1.81	4.80	4.17	3.40	2.40
16	5/8	2.99	2.24	5.99	5.17	4.24	2.99
19	3/4	4.31	3.22	8.62	7.48	6.08	4.31
22	7/8	5.80	4.35	11.61	10.07	8.21	5.80
26	1	7.57	5.67	15.15	13.11	10.70	7.57
28	1⅛	9.61	7.21	19.23	16.64	13.60	9.61
32	1¼	11.88	8.93	23.76	20.59	16.78	11.88
35	1⅜	14.69	11.02	29.39	25.44	20.77	14.69
38	1½	17.41	13.06	34.83	30.16	24.63	17.41
40	1⅝	20.50	15.37	41.00	35.51	28.98	20.50
44	1¾	23.58	17.69	47.16	40.81	33.33	23.58
48	1⅞	27.57	20.68	55.15	47.75	39.00	27.57
52	2	30.66	22.99	61.31	53.10	43.35	30.66
56	2¼	38.09	28.57	76.19	65.98	53.88	38.09
64	2½	47.16	35.37	94.33	81.68	66.66	47.16
69	2¾	55.32	41.50	110.65	95.82	78.23	55.32

If used with Double Choker Hitch multiply above values by ¾

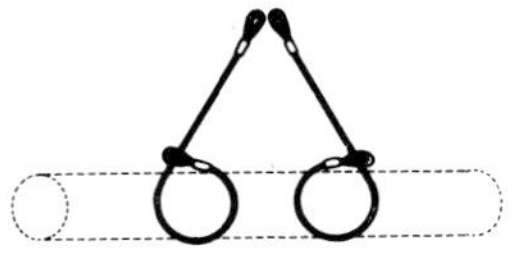

For Double Basket Hitch multiply above values by 2.

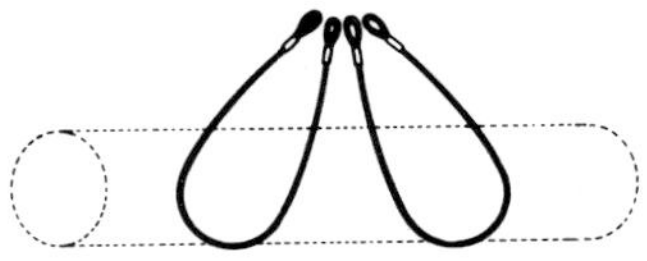

Note: Table values are for slings with eyes and thimbles in both ends, Flemish spliced eyes and mechanical sleeves.
Hand-tucked spliced eyes – reduce loads according to Table 1.11.
Eyes formed by cable clips – reduce loads by 20%.
See also BS.1290 (Wire Rope Slings and Sling Legs).

SLINGS

TABLE 6.12
(Canadian Standard)

WIRE ROPE SLINGS
6 × Classification Group, Improved Plow Steel (Grade 160) IWRC

MAXIMUM SAFE WORKING LOADS (SWL) – tonnef
(Safety Factor = 5)

Rope diameter		Single vertical hitch	Single choker hitch	Single basket hitch (vertical legs)	Two-leg bridle hitch and single basket hitch with legs inclined		
mm	in				60°	45°	30°
6	¼	0.52	0.39	1.04	0.91	0.73	0.52
8	5/16	0.79	0.59	1.59	1.36	1.13	0.79
9	⅜	1.16	0.86	2.31	2.00	1.63	1.16
11	7/16	1.56	1.18	3.13	2.72	2.22	1.56
13	½	2.13	1.59	4.26	3.70	3.02	2.13
14	9/16	2.58	1.90	5.17	4.49	3.65	2.58
16	⅝	3.22	2.40	6.44	5.58	4.53	3.22
19	¾	4.63	3.47	9.25	8.03	6.53	4.63
22	⅞	6.24	4.67	12.47	10.79	8.80	6.24
26	1	8.14	6.10	16.28	14.10	11.52	8.14
28	1⅛	10.32	7.71	20.63	17.86	14.60	10.32
32	1¼	12.79	9.61	25.58	22.13	18.09	12.79
35	1⅜	15.78	11.83	31.56	27.35	22.31	15.78
38	1½	18.73	14.06	37.46	32.43	26.48	18.73
40	1⅝	22.04	16.51	44.08	38.18	31.16	22.04
44	1¾	23.35	19.00	50.70	43.90	35.83	23.35
48	1⅞	29.65	22.22	59.32	51.38	41.95	29.65
52	2	32.92	24.72	65.85	57.00	46.57	32.92
56	2¼	40.95	30.66	81.90	70.93	57.91	40.95
64	2½	50.70	37.96	101.40	87.80	71.70	50.70
69	2¾	59.45	44.53	118.91	102.94	84.08	59.45

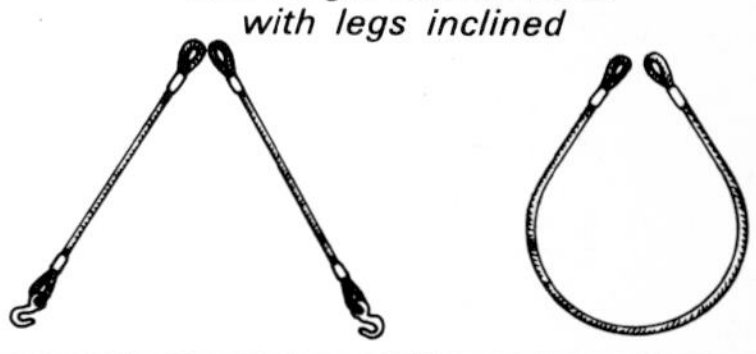

If used with Double Choker Hitch multiply above values by ¾

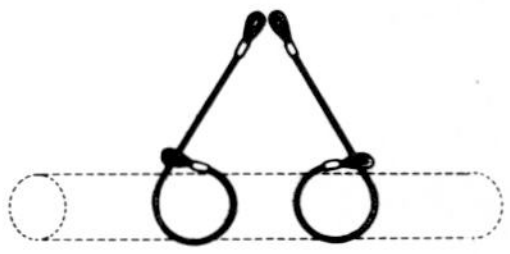

For Double Basket Hitch multiply above values by 2.

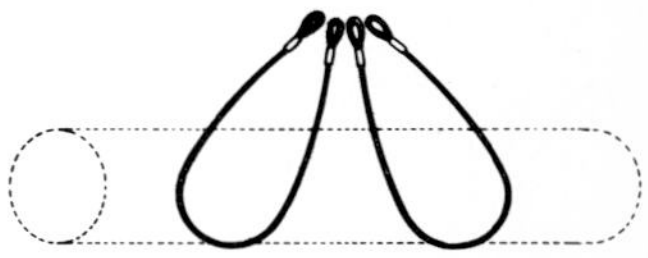

Note: Table values are for slings with eyes and thimbles in both ends, Flemish spliced eyes and mechanical sleeves.
Hand-tucked spliced eyes – reduce loads according to Table 1.11.
Eyes formed by cable clips – reduce loads by 20%.
See also BS.1290 (Wire Rope Slings and Sling Legs).

TABLE 6.13
(Canadian Standard)

WIRE ROPE SLINGS
6 × 37 Classification Group, Improved Plow Steel (Grade 160) Fibre Core

MAXIMUM SAFE WORKING LOADS (SWL) – tonnef
(Safety Factor = 5)

Rope diameter		Single vertical hitch	Single choker hitch	Single basket hitch (vertical legs)	Two-leg bridle hitch and single basket hitch with legs inclined		
mm	in				60°	45°	30°
6	¼	0.45	0.34	0.91	0.79	0.63	0.45
8	5/16	0.73	0.54	1.45	1.25	1.02	0.73
9	3/8	0.98	0.75	2.00	1.72	1.40	0.98
11	7/16	1.36	1.02	2.72	2.36	1.93	1.36
13	½	1.81	1.36	3.63	3.13	2.56	1.81
14	9/16	2.27	1.70	4.54	3.92	3.22	2.27
16	5/8	2.90	2.18	5.80	5.03	4.10	2.90
19	¾	4.04	3.04	8.07	6.98	5.71	4.04
22	7/8	5.49	4.13	10.97	9.52	7.75	5.49
26	1	7.17	5.40	14.33	12.43	10.11	7.17
28	1/8	8.89	6.67	17.78	15.37	12.56	8.89
32	1¼	11.07	8.30	22.13	19.18	15.65	11.07
35	1 3/8	13.51	10.16	27.03	23.40	19.09	13.51
38	1½	16.33	12.24	32.65	28.30	23.08	16.33
40	1 5/8	19.14	14.38	38.28	33.15	27.07	19.14
44	1¾	21.95	16.46	43.90	38.00	31.01	21.95
48	1 7/8	25.76	19.32	51.52	44.62	36.42	25.76
52	2	28.12	21.09	56.23	48.70	39.77	28.12
56	2¼	36.46	27.35	72.92	63.17	51.56	36.46
64	2½	44.44	33.33	88.89	76.96	62.86	44.44
69	2¾	53.15	39.86	106.30	92.06	75.14	53.15

If used with Double Choker Hitch multiply above values by ¾

For Double Basket Hitch multiply above values by 2.

Note: Table values are for slings with eyes and thimbles in both ends, Flemish spliced eyes and mechanical sleeves. Hand-tucked spliced eyes – reduce loads according to Table 1.11.
Eyes formed by cable clips – reduce loads by 20%.
See also BS.1290 (Wire Rope Slings and Sling Legs).

SLINGS

TABLE 6.14
(Canadian Standard)

WIRE ROPE SLINGS
6 × 37 Classification Group, Improved Plow Steel (Grade 160) IWRC

MAXIMUM SAFE WORKING LOADS (SWL) – tonnef
(Safety Factor = 5)

Rope diameter		Single vertical hitch	Single choker hitch	Single basket hitch (vertical legs)	Two-leg bridle hitch and single basket hitch with legs inclined		
mm	in				60°	45°	30°
6	1/4	0.48	0.36	0.95	0.82	0.68	0.48
8	5/16	0.77	0.59	1.54	1.34	1.09	0.77
9	3/8	1.07	0.79	2.13	1.86	1.50	1.07
11	7/16	1.45	1.09	2.90	2.52	2.04	1.45
13	1/2	1.95	1.45	3.90	3.38	2.77	1.95
14	9/16	2.43	1.81	4.85	4.19	3.42	2.43
16	5/8	3.13	2.36	6.26	5.42	4.42	3.13
19	3/4	4.31	3.22	8.62	7.46	6.08	4.31
22	7/8	5.90	4.42	11.79	10.20	8.34	5.90
26	1	7.71	5.78	15.42	13.36	10.88	7.71
28	1 1/8	9.52	7.14	19.05	16.50	13.47	9.52
32	1 1/4	11.88	8.91	23.76	20.59	16.78	11.88
35	1 3/8	14.51	10.88	29.02	25.12	20.50	14.51
38	1 1/2	17.91	13.42	35.83	31.02	25.35	17.91
40	1 5/8	20.59	15.42	41.18	35.65	29.11	20.59
44	1 3/4	23.58	17.69	47.16	40.82	33.33	23.58
48	1 7/8	27.66	20.75	55.32	47.93	39.14	27.66
52	2	30.20	22.65	60.41	52.33	42.72	30.20
56	2 1/4	39.18	29.39	78.36	67.84	55.42	39.18
64	2 1/2	47.75	35.83	95.51	82.72	67.53	47.75
69	2 3/4	57.14	42.86	114.28	98.95	80.81	57.14

If used with Double Choker Hitch multiply above values by 3/4

For Double Basket Hitch multiply above values by 2.

Note: Table values are for slings with eyes and thimbles in both ends, Flemish spliced eyes and mechanical sleeves. Hand-tucked spliced eyes – reduce loads according to Table 1.11.
Eyes formed by cable clips – reduce loads by 20%.
See also BS.1290 (Wire Rope Slings and Sling Legs).

TABLE 6.15
(Canadian Standard)

BRAIDED WIRE ROPE SLINGS
6-Part Braided Rope
6 × 19 Classification Group, Improved Plow Steel (Grade 160)

MAXIMUM SAFE WORKING LOADS (SWL) – tonnef
(Safety Factor = 5)

Component rope diameter		Single vertical hitch	Single choker hitch	Single basket hitch (vertical legs)	Two-leg bridle hitch and single basket hitch with legs inclined		
mm	in				60°	45°	30°
5	3/16	1.18	0.88	2.36	2.04	1.68	1.18
6	1/4	2.09	1.56	4.17	3.61	2.95	2.09
8	5/16	3.27	2.45	6.53	5.67	4.63	3.27
9	3/8	4.63	3.47	9.25	8.03	6.53	4.63
11	7/16	6.26	4.72	12.52	10.84	8.84	6.26
13	1/2	8.16	6.12	16.33	14.15	11.56	8.16
14	9/16	9.98	7.48	19.95	17.28	14.10	9.98
16	5/8	12.70	9.52	25.40	21.90	17.96	12.70
19	3/4	18.14	13.60	36.28	31.43	25.67	18.14
22	7/8	24.49	18.37	48.98	42.40	34.65	24.49
26	1	31.75	23.81	63.49	54.96	44.90	31.75

If used with Double Choker Hitch multiply above values by ¾

For Double Basket Hitch multiply above values by 2.

Note: Table values are for braided slings with eyes and thimbles in both ends. Eyes formed by hand tucking ropes and securing with mechanical sleeves.
See also BS.1290 (Wire Rope Slings and Sling Legs).

SLINGS

TABLE 6.16
(Canadian Standard)

BRAIDED WIRE ROPE SLINGS
8-Part Braided Rope
6 × 19 Classification Group, Improved Plow Steel (Grade 160)

MAXIMUM SAFE WORKING LOADS (SWL) – tonnef
(Safety Factor = 5)

Component rope diameter		Single vertical hitch	Single choker hitch	Single basket hitch (vertical legs)	Two-leg bridle hitch and single basket hitch with legs inclined		
mm	in				60°	45°	30°
5	3/16	1.54	1.16	3.08	2.68	2.18	1.54
6	1/4	2.81	2.11	5.62	4.85	3.97	2.81
8	5/16	4.35	3.27	8.17	7.53	6.17	4.35
9	3/8	6.17	4.63	12.34	10.70	8.71	6.17
11	7/16	8.44	6.35	16.87	14.60	11.93	8.44
13	1/2	10.88	8.16	21.77	18.37	15.37	10.88
14	9/16	13.60	10.20	27.21	23.58	19.23	13.60
16	5/8	17.23	12.92	34.47	29.84	24.35	17.23
19	3/4	24.49	18.37	48.98	42.40	34.65	24.49
22	7/8	32.65	24.49	65.30	56.55	46.17	32.65
26	1	42.63	31.97	85.26	73.83	60.27	42.63

If used with Double Choker Hitch multiply above values by 3/4

For Double Basket Hitch multiply above values by 2.

Note: Table values are for braided slings with eyes and thimbles in both ends. Eyes formed by hand tucking ropes and securing with mechanical sleeves.
See also BS.1290 (Wire Rope Slings and Sling Legs).

SLINGS

TABLE 6.17
(Canadian Standard)

CABLE-LAID WIRE ROPE SLINGS
Mechanical Splices

MAXIMUM SAFE WORKING LOADS (SWL) – tonnef
(Safety Factor = 5)

Cable body diameter		Single vertical hitch	Single choker hitch	Single basket hitch (vertical legs)	Two-leg bridle hitch and single basket hitch with legs inclined		
mm	in				60°	45°	30°
6	1/4	0.45	0.34	0.90	0.79	0.63	0.45
9	3/8	1.00	0.75	2.00	1.72	1.41	1.00
13	1/2	1.63	1.22	3.27	2.83	2.31	1.63
16	5/8	2.54	1.90	5.08	4.40	3.58	2.54
19	3/4	3.45	2.58	6.89	5.99	4.85	3.45
22	7/8	4.54	3.40	9.07	7.85	6.39	4.54
26	1	5.80	4.35	11.61	10.07	8.21	5.80
28	1 1/8	6.98	5.26	13.97	12.11	9.89	6.98
32	1 1/4	8.34	6.26	16.69	14.47	11.79	8.34
35	1 3/8	9.98	7.48	19.95	17.28	14.10	9.98
38	1 1/2	11.79	8.84	23.58	20.41	16.69	11.79

If used with Double Choker Hitch multiply above values by 3/4

For Double Basket Hitch multiply above values by 2.

Note: Cable-laid wire rope slings with mechanical splices are slings made from cable and having eyes fabricated by pressing or swaging one or more metal sleeves over the rope junction at the eyes. The cable consists of six individual wire ropes wrapped around a seventh wire rope which forms the core. For cable body diameters of 6–16 mm (1/4–5/8 in) the construction is 7 × 7 × 7 (i.e., 7 wire ropes of 7 × 7 construction) and for diameters above 16 mm (5/8 in) it is 7 × 6 × 19 IWRC (i.e., 7 wire ropes of 6 × 19 IWRC construction).
The table values are for slings with eyes and thimbles in both ends.
See also BS.1290 (Wire Rope Slings and Sling Legs).

SLINGS

TABLE 6.18
(Canadian Standard)

STRAND-LAID GROMMET SLINGS
Hand Tucked, Improved Plow Steel (Grade 160)

MAXIMUM SAFE WORKING LOADS (SWL) – tonnef
(Safety Factor = 5)

Rope body diameter		Single vertical hitch	Single choker hitch	Single basket hitch (vertical legs)	Single basket hitch with legs inclined		
mm	in				60°	45°	30°
6	1/4	0.77	0.58	1.54	1.34	1.09	0.77
8	5/16	1.18	0.88	2.36	2.04	1.68	1.18
9	3/8	1.72	1.29	3.45	2.99	2.45	1.72
11	7/16	2.36	1.77	4.72	4.08	3.33	2.36
13	1/2	3.17	2.38	6.35	5.49	4.49	3.17
14	9/16	3.81	2.86	7.62	6.58	5.40	3.81
16	5/8	4.72	3.54	9.43	8.16	6.67	4.72
19	3/4	6.71	5.03	13.42	11.61	9.48	6.71
22	7/8	9.07	6.80	18.14	15.69	12.83	9.07
26	1	11.79	8.84	23.58	20.40	16.69	11.79
28	1 1/8	13.51	10.16	27.03	23.40	19.09	13.51
32	1 1/4	15.87	11.93	31.75	27.48	22.45	15.87
35	1 3/8	19.32	14.51	38.64	33.47	27.30	19.32
38	1 1/2	23.40	17.55	46.80	40.54	33.11	23.40
40	1 5/8	25.67	19.27	51.34	44.44	36.28	25.67
44	1 3/4	29.48	22.13	58.96	51.06	41.68	29.48
48	1 7/8	34.47	25.85	68.93	59.68	48.75	34.47
52	2	37.64	28.25	75.28	65.21	53.24	37.64
56	2 1/4	45.80	34.38	91.60	79.32	64.76	45.80

For Double Basket Hitch multiply above values by 2.

Note: A strand-laid grommet is an endless wire rope sling fabricated from a single strand that is wrapped six times. After wrapping, the two free strand ends are tucked inside the wraps to form the core of the grommet. Grommet diameters up to and including 26 mm (1 in) are fabricated from strands of 6 × 19 rope and grommet diameters of 28 mm (1 1/8 in) and greater are fabricated from strands of 6 × 37 rope.
See also BS.1290 (Wire Rope Slings and Sling Legs).

TABLE 6.19
(Canadian Standard)

CABLE-LAID GROMMET SLINGS
Hand Tucked, Improved Plow Steel (Grade 160)

MAXIMUM SAFE WORKING LOADS (SWL) – tonnef
(Safety Factor = 5)

Cable body diameter		Single vertical hitch	Single choker hitch	Single basket hitch (vertical legs)	Single basket hitch with legs inclined		
mm	in				60°	45°	30°
9	3/8	1.18	0.88	2.36	2.04	1.68	1.18
14	9/16	2.54	1.90	5.08	4.40	3.58	2.54
16	5/8	3.54	2.65	7.07	6.12	4.99	3.54
19	3/4	4.63	3.47	9.25	8.03	6.53	4.63
24	15/16	7.17	5.40	14.33	12.43	10.11	7.17
28	1 1/8	9.98	7.48	19.95	17.28	14.10	9.98
33	1 5/16	13.60	10.20	27.21	23.58	19.33	13.60
38	1 1/2	17.23	12.92	34.47	29.84	24.35	17.23
43	1 11/16	21.77	16.33	43.54	37.69	30.79	21.77
48	1 7/8	27.21	20.41	54.42	47.12	38.46	27.21
56	2 1/4	38.09	28.57	76.19	65.98	53.88	38.09
60	2 5/8	50.79	38.09	101.58	87.98	71.83	50.79

For Double Basket Hitch multiply above values by 2.

Note: A cable-laid, hand-tucked grommet is an endless wire rope sling fabricated from a single wire rope that is wrapped six times. After wrapping, the two free ends of the rope are tucked inside the wraps to form the core of the grommet. It is very similar to the strand-laid grommet except that a complete rope rather than a strand is used to form the grommet. For cable body diameters of 9 and 14 mm (3/8 and 9/16 in) the construction is 7 × 6 × 7 (the first figure denotes six rope wraps plus one rope core, and the second two digits indicate that the rope used is 6 × 7). For diameters of 16 mm (5/8 in) and up the construction is 7 × 6 × 19 (i.e., six rope wraps plus one rope core, all of 6 × 19 rope).
See also BS.1290 (Wire Rope Slings and Sling Legs).

SLINGS

TABLE 6.20
(Canadian Standard)

STRAND-LAID ENDLESS SLINGS
Mechanical Splice, Improved Plow Steel (Grade 160)

MAXIMUM SAFE WORKING LOADS (SWL) – tonnef
(Safety Factor = 5)

Rope body diameter		Single vertical hitch 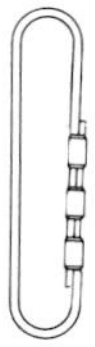	Single choker hitch 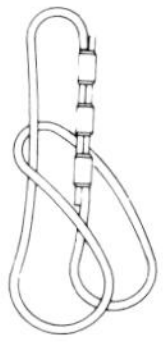	Single basket hitch (vertical legs) 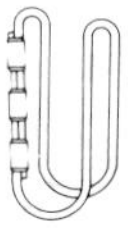	Single basket hitch with legs inclined 		
mm	in				60°	45°	30°
6	¼	0.84	0.63	1.68	1.45	1.18	0.84
9	⅜	1.81	1.36	3.63	3.13	2.26	1.81
13	½	3.27	2.45	6.53	5.67	4.63	3.27
16	⅝	5.08	3.80	10.16	8.80	7.17	5.08
19	¾	7.26	5.44	14.51	12.56	10.25	7.26
22	⅞	9.98	7.48	19.95	17.28	14.10	9.98
26	1	12.70	9.52	25.40	21.99	17.96	12.70
28	1⅛	16.33	12.24	32.65	28.30	23.08	16.33
32	1¼	19.05	14.29	38.09	32.97	26.94	19.05
35	1⅜	22.68	17.00	45.35	39.27	32.06	22.68
38	1½	26.30	19.73	52.60	45.58	37.19	26.30

For Double Basket Hitch multiply above values by 2.

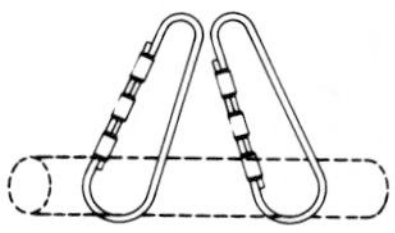

Note: A strand-laid endless sling with mechanical splice is a sling fabricated from a single wire rope that is bent into a loop and whose ends are attached to each other by means of one or more mechanical fittings. Fore rope diameters up to and including 28 mm (1⅛ in) the rope used is 6 × 19 IWRC and for diameters greater than 28 mm (1⅛ in) it is 6 × 37 IWRC.

See also BS.1290 (Wire Rope Slings and Sling Legs).

TABLE 6.21
(Canadian Standard)

Cable body diameter		Single vertical hitch 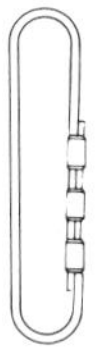	Single choker hitch 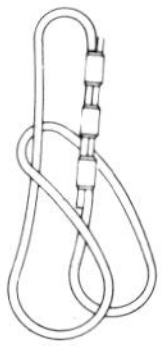	Single basket hitch (vertical legs) 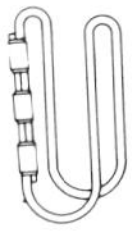	Single basket hitch with legs inclined 		
					MAXIMUM SAFE WORKING LOADS (SWL) – tonnef (Safety Factor = 5)		
mm	in				60°	45°	30°
6	¼	0.75	0.57	1.50	1.29	1.06	0.75
9	⅜	1.63	1.22	3.26	2.83	2.31	1.63
13	½	2.72	2.04	5.44	4.71	3.85	2.72
16	⅝	4.08	3.06	8.16	7.07	5.76	4.08
19	¾	5.62	4.22	11.25	9.75	7.94	5.62
22	⅞	7.53	5.67	15.06	13.06	10.66	7.53
26	1	9.07	6.80	18.14	15.69	12.83	9.07
28	1⅛	11.79	8.84	23.58	20.40	16.69	11.78
32	1¼	14.51	10.88	29.02	25.12	20.50	14.51
35	1⅜	16.32	12.24	32.65	28.30	23.08	16.32
38	1½	19.95	14.97	39.90	34.56	28.21	19.95

CABLE-LAID ENDLESS SLINGS
Mechanical Splice, Improved Plow Steel (Grade 160)

For Double Basket Hitch multiply above values by 2.

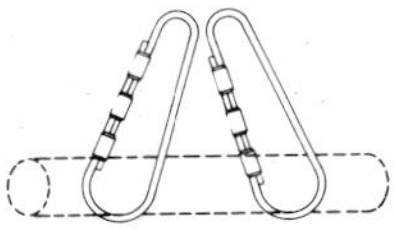

Note: Cable-laid endless slings with mechanical splices are endless slings fabricated from cable that is bent into a loop and whose ends are attached to each other by means of one or more mechanical fittings. The cable is composed of six individual wire ropes wrapped around a seventh wire rope which forms the core. For cable body diameters of 6–16 mm (¼–⅝ in) the construction is 7 × 7 × 7 (i.e., 7 wire ropes of 7 × 7 construction) and for diameters above 16 mm (⅝ in) it is 7 × 6 × 19 IWRC (i.e., 7 wire ropes of 6 × 19 IWRC construction).

See also BS.1290 (Wire Rope Slings and Sling Legs).

CHAPTER 7

Operating procedures and precautions

RESPONSIBILITIES

It is not only the men on the job who have responsibilities for their own and their fellow workers' safety. Management also has responsibilities that must be met.

It is the responsibility of management and supervision to ensure that the men who prepare the equipment, use the equipment and work with or around it are well trained in both safety and operating procedures.

The employer must ensure that all lifting equipment is operated only by trained, experienced and competent operators.

The employer must also ensure that the men who direct, rig and handle the loads have received training in the principles of the operation, are able to establish weights and judge distances, heights and clearances, are capable of selecting tackle and lifting gear suitable for the loads to be lifted, and are capable of directing the movement of the crane and load to ensure the safety of all personnel.

The employer is also responsible for putting together a safety programme, educating all personnel in safe practices, and the assignment to the lifting crews of definite, individual safety responsibilities.

These responsibilities must be assigned by management to on-the-job personnel. Job titles may vary, but the essential responsibilities can be allocated as follows:

(a) *Planning:* Major lifting operations must be *planned* and supervised by competent personnel to ensure that the best methods and most suitable equipment and tackle are employed.

(b) *Supply and care of lifting equipment:* Job management must ensure that:
  – Correct lifting equipment is available.
  – Correct load ratings are available for the material and equipment used for lifting.
  – Lifting material and equipment are maintained in efficient working condition.

(c) *Lifting operation:* The supervisor of the lifting operation should be responsible for:
  – Correct suspension of the load.
  – Supervision of the lifting crew.
  – Ensuring that the lifting material and equipment have the necessary capacity for the job and are in safe condition.
  – Ensuring correct assembly of lifting material or equipment as required during the operation, such as the correct installation of lifting bolts.
  – Safety of the lifting crew and other personnel as they are affected by the lifting operation.

PROCEDURES AND PRECAUTIONS

The single most important precaution in a lifting operation is to determine the weight of all loads before attempting to lift them, making ample allowances for unknown factors, and determining the available capacity of the equipment being used. In cases where the

OPERATING PROCEDURES AND PRECAUTIONS

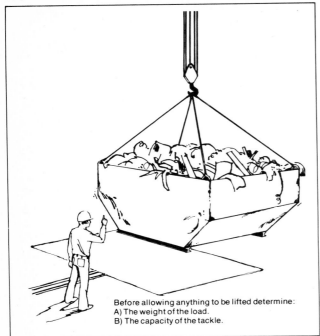

Before allowing anything to be lifted determine:
A) The weight of the load.
B) The capacity of the tackle.

Fig. 7.1 Know the load weight and tackle capacity

Destroy defective components before throwing them away.

They could be used by someone not aware of the hazards or defects.

Fig. 7.2 Destroy defective components

assessment of load weight is difficult, safe load indicators or weighing devices should be fitted. This chapter also includes a section dealing with the estimation of load weights.

It is also important to lift the load only when it is stable. Unless the centre of gravity of the load is below the hook, the load will shift. A separate section dealing with centres of gravity has been included in this chapter.

The safety of personnel involved in lifting operations depends largely upon care and common sense. Remember these safe practices:

– Know the safe working load (SWL) of the equipment and tackle being used. Never exceed this limit. (Fig. 7.1)
– Determine the load weight before slinging and lifting it.
– Examine all equipment, tackle and slings before using them and *destroy* defective components. Discarded equipment may be used by someone not aware of the hazards or defects. (Fig. 7.2)
– No one who has reasonable cause to believe that any equipment or tackle which has been assigned to him is unsafe or unsuited to the job should use or operate it until he has reported the defect or hazard to his supervisor, safe conditions have been assured, and orders to proceed have been issued by some person in authority who is then responsible for the safety of all personnel exposed to the unsafe conditions.
– Never carry out any lifting operations when the weather conditions are such that hazards to personnel, property or the public are created. The size and shape of the loads being lifted must be carefully assessed to determine if a safety hazard exists during high wind speeds. Avoid handling loads presenting large wind catching surfaces which could result in loss of control of the load during times of high or gusty winds, even though the weight of the load is within the normal capacity of the equipment. Wind loading can be critical for the manner in which the load is landed and the safety of the men handling it. When winds reach about 40–50 km/h (25–30 mph) consider limiting operations. (Fig. 7.3)
– If the visibility of the riggers or lifting crew is impaired by dust, darkness, snow, fog or rain, strict supervision of the operation must be exercised and, if necessary, it should be suspended. (Fig. 7.4)

OPERATING PROCEDURES AND PRECAUTIONS

Fig. 7.3 Wind can severely affect a load

Fig. 7.4 Adequate lighting should be provided for all night-time operations

OPERATING PROCEDURES AND PRECAUTIONS

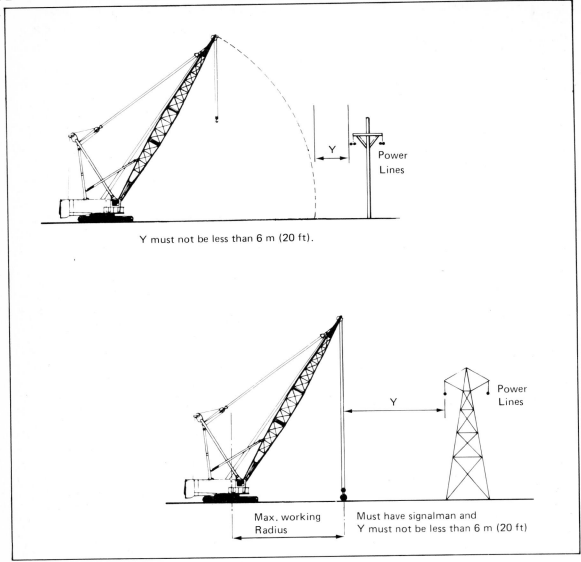

Fig. 7.5 Requirements when working near power lines

— Whenever the temperature is below freezing (0°C, 32°F), extreme caution must be exercised to ensure that no part of the hoist or crane structure or tackle is shock loaded or impacted as brittle fracture of the steel can result.
— **The most repeated killer of crane crews and those persons handling loads is electrocution caused by the contact of the jib hoist rope or load of a crane with electric power lines.** When working with or around cranes that are within a jib's length of any power line ensure that a competent signalman is stationed at all times within view of the operator to warn him when any part of the machine or its load is approaching the minimum safe distance (6 m) from the power line. Caution must also be exercised when working near overhead lines having long spans as they tend to swing laterally due to the wind and accidental contact could occur. (Fig. 7.5)

- The SWL of all lifting and rigging equipment and tackle are based on almost ideal conditions seldom achieved under actual working conditions, and as such it is important to recognise the factors which can reduce the capacity of the equipment.
- The SWL of lifting equipment apply only to freely-suspended loads on plumb ropes. If the lift line is not plumb at all times when handling loads then additional side loads will hazard the stability of the equipment and introduce stresses for which it has not been designed. In circumstances such as this, structural failures can occur without any warning. (Fig. 7.6)

Fig. 7.6 All lifting must be carried out with lines plumb

- Rapid slewing of suspended loads also subjects the equipment to additional stresses which can cause collapse. The force of the slewing action causes the load to drift away from the machine, thus increasing the radius and side loading the equipment. The load must always be kept directly below the jib head or upper load block. (Fig. 7.7)
- The SWL also apply only to equipment in good condition, having undamaged and unkinked structural members. If any equipment becomes damaged it should be taken out of service until the necessary repairs have been effected.
- The maximum SWL of most rigging and lifting equipment are determined from static loads and the safety factor is applied to account for dynamic motions of the load and equipment. In order to ensure that the SWL is not exceeded during operation, allowances should be made for wind loading and dynamic forces set up by the normal operational movement of the machine and load. It is essential to avoid sudden snatching, slewing and stopping of suspended loads since rapid acceleration and deceleration will greatly increase the stresses in the equipment and tackle.
- The rated loads of most lifting equipment do not generally account for the weight of hook blocks, hooks, slings, equaliser beams, material handling equipment and other elements of lifting tackle. Their combined weight must be subtracted from the load capacity of the equipment to determine the maximum allowable load to be lifted. (Fig. 7.8)

The life and safety of slings and their contribution to the security of lifting operations can be greatly increased by taking care in their application.

- Never use kinked or damaged slings or lift ropes. To provide maximum operating efficiency and safety, all slings and fittings should be given thorough periodic examinations as well as daily inspections for signs of wear and abrasion, broken wires, worn or cracked fittings, loose seizings and splices, kinking, crushing, flattening, and corrosion. Special care should be taken in inspecting the areas around thimbles and fittings.
- All slings should be identified with an identification number and their maximum capacity on a flat ferrule or ring, permanently attached to the sling. Mark the capacity of the sling for a vertical load or at an angle of 45°, but be certain that all employees are aware of how the rating system works. (Fig. 7.9)
- Sharp bends, pinching and crushing must be avoided. Loops and thimbles should be used at all times. Corner pads that prevent the sling from being sharply bent or cut can consist of large diameter split pipe sections, corner saddles, padding or blocking. (Fig. 7.10)

It's a good rule to make sure that the length of the arc of contact of the rope is at least equal to one rope lay (about seven times the rope diameter). When the bend is of this length, each of the strands has been on the inner and outer sides of the rope bend and the slippage of the strands relative to each other minimises the stress. On the other hand, if the bend is very short, the strands on the outer

OPERATING PROCEDURES AND PRECAUTIONS

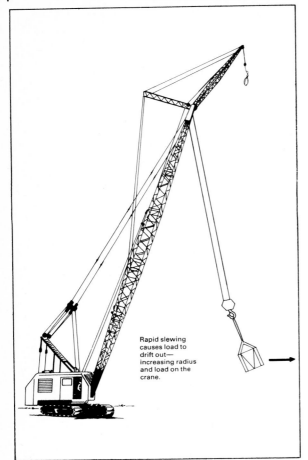

Fig. 7.7 Keep the load under control at all times

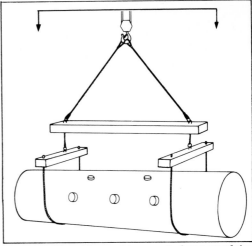

Fig. 7.8 All tackle must be counted as part of the load

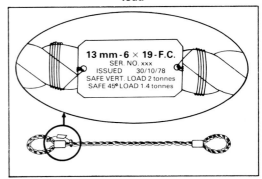

Fig. 7.9 All slings should carry identification tags

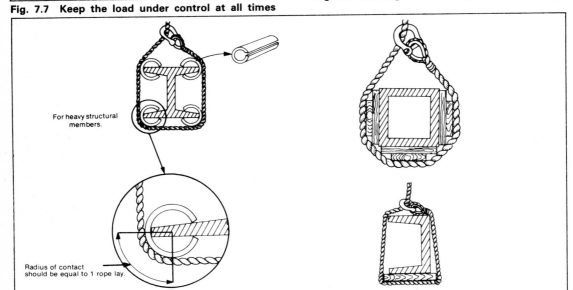

Fig. 7.10 Protect slings at all sharp corners on heavy items

OPERATING PROCEDURES AND PRECAUTIONS

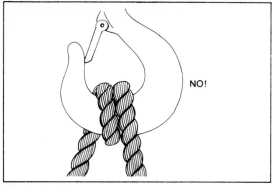

Fig. 7.11 Never wrap rope round a hook

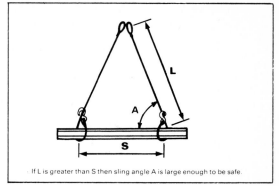

Fig. 7.13 Check sling angle

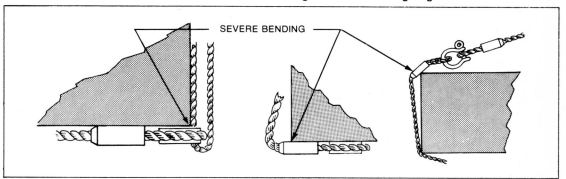

Fig. 7.12 Avoid bending rope near any splice or attached fitting

side of the rope bend will have to stretch and this stretch will usually be permanent and will leave a sharp bend or kink in the sling.

- Never allow wire rope to lie on the ground for any length of time or on damp or wet surfaces, rusty steel or near corrosive substances.
- Avoid dragging rope slings from beneath loads.
- Keep all rope away from flame cutting and electric welding operations.
- Avoid contact with solvents and chemicals.
- Knotted and kinked slings are permanently damaged and should not be used.
- Never use discarded hoist rope as sling material.
- Avoid using single-leg wire rope slings with hand-spliced eyes. The load can spin, thus causing the rope to unlay and allow the splice to pull out. Slings having Flemish spliced eyes should be used.
- Never wrap a wire rope completely around a hook. The sharp radius will damage the rope. (Fig. 7.11)

- Avoid bending the eye section of wire rope slings around corners. The bend will weaken the splice or swaging. There must be no bending near any attached fitting. (Fig. 7.12)
- Ensure that the sling angle to the horizontal is always greater than 45°. To make sure that the angle is adequate once a load is slung, check that the horizontal distance between the attachment points on the load is less than the length of the shortest sling leg. If this is the case then the sling angle is greater than 60°. (Fig. 7.13)
- Do not assume that a multi-legged bridle sling will safely lift a load equal to the safe load on one leg multiplied by the number of legs. There is no way of knowing that each leg is carrying its share of the load. With slings having more than two legs and a rigid load, it is possible for two of the legs to take practically the full load while the others only balance it. Due consideration with regard to this has been given in computing the SWL tables for three- and four-legged bridle slings which are given only the same safe load as a two-legged sling. (Fig. 7.14)

OPERATING PROCEDURES AND PRECAUTIONS

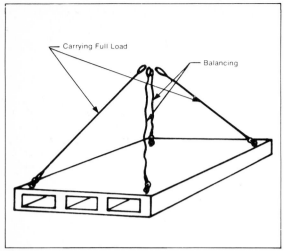

Fig. 7.14 The load of a rigid object could be carried by only two legs of a multi-leg bridle sling while the others serve only to balance

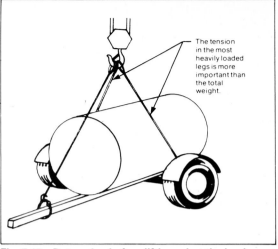

Fig. 7.15 Determine before lifting what the load on each sling leg will be

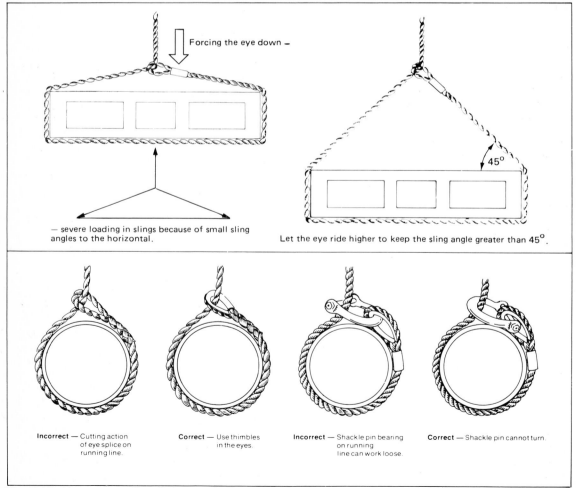

Fig. 7.16 Use correct sling fittings and allow suitable sling angle

OPERATING PROCEDURES AND PRECAUTIONS

- When lifting rigid objects with slings having three or four legs, any two of the legs must be capable of supporting the total load. In other words, after assessing the angle between the legs, the slings should be considered as having only two legs when estimating the size required. Where the object is flexible and able to bend to adjust itself to the sling legs, each leg can be assumed to take its own share of the load.
- When using multi-legged slings to lift loads in which one end is much heavier than the other the tension in the most heavily loaded leg is much more important than the total weight. The sling must be selected to suit the most heavily loaded leg rather than the total weight. (Fig. 7.15)
- When using choker hitches, do not force the eye down towards the load once tension is applied. Rope damage is the invariable result. (Fig. 7.16)
- Whenever two or more rope eyes must be placed over a hook, install a shackle on the hook with the shackle pin resting in the hook and place the rope eyes on the shackle. This will prevent the spread of the sling legs from opening up the hook and also prevent the eyes from damaging each other when under load. (Fig. 7.17)

The following procedures and precautions should be observed whenever loads are to be handled:

- All loads must be correctly secured to prevent the dislodgement of any part. Suspended loads should be securely slung and properly balanced before they are set in motion.
- The load must be kept under control at all times. Where necessary, when personnel may be endangered by the rotation, one or more guide ropes or tag lines should be used to prevent the rotation or uncontrolled motion. (Fig. 7.18)
- Loads must be safely landed and properly blocked before being unhooked and unslung. (Fig. 7.19)
- Lifting beams should be plainly marked with their weight and designed working loads and should be used only for the purpose for which they were designed.
- The lifting rope must never be wrapped around the load. The load should be attached to the hook by slings or other tackle devices that are adequate for the load being lifted.
- Multi-part lines must not be twisted around each other.
- The load line should be brought over the centre of gravity of load before the lift is started.
- If there has been a slack rope condition, determine that the rope is properly seated on the drum and in the sheaves.
- Materials and equipment being hoisted must be loaded and secured to prevent any movement which could create a hazard in transit. (Fig. 7.20)
- Keep hands away from pinch points as the slack is being taken up.
- Wear gloves when handling wire rope.
- Make sure that all personnel stand clear while loads are being lifted and lowered or while the slings are being drawn from beneath the load. The hooks may catch under the load and suddenly fly free. (Fig. 7.21)
- Before making a lift, check to see that the sling is properly attached to the load.
- Avoid impact loading caused by sudden jerking when lifting or lowering. Lift the sling gradually until the slack is eliminated.
- Never ride on a load that is being lifted.
- Never allow the load to be carried over the heads of any personnel.
- Never work under a suspended load unless the load has been adequately supported from the floor and all conditions have been approved by the supervisor in charge of the operation.
- Never leave a load suspended in the air when the hoist or crane is unattended.
- Never make temporary repairs to a sling. Procedures for proper repairs to a damaged sling should be established and followed.
- Never lift loads with one leg of a multi-leg sling until the unused legs are made secure. (Fig. 7.22)
- Never point load a hook unless it is especially designed and rated for such use.
- Remove or secure all loose pieces of material from the load before it is moved.
- Make sure that the load is free before lifting and that all sling legs are taking the load.
- When using two or more slings on a load ensure that all slings are made from the same materials.
- Lower the loads on to adequate blocking to prevent damage to the slings.

OPERATING PROCEDURES AND PRECAUTIONS

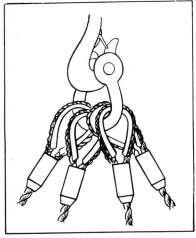

Fig. 7.17 When two or more ropes are needed to lift, use a shackle on the hook

Fig. 7.18 Use tag lines to control all loads

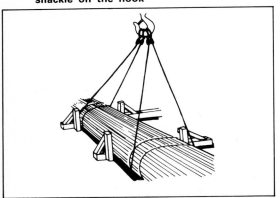

Fig. 7.19 Before unhooking loads ensure that they are safely landed and suitably blocked

Fig. 7.20 Load all materials securely to prevent their moving or dislodging

Fig. 7.21 Keep clear of slings being pulled from under loads

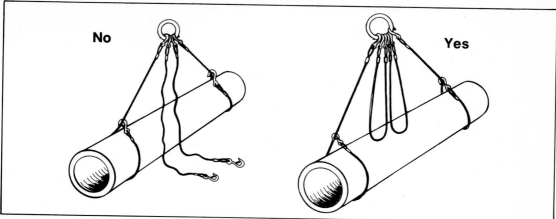

Fig. 7.22 Secure all unused sling legs

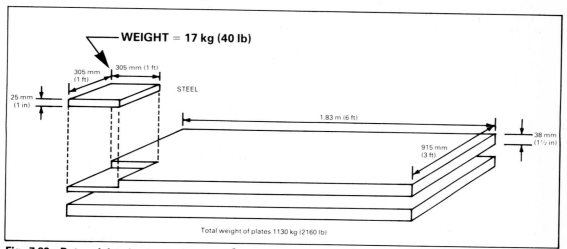

Fig. 7.23 Determining load weight (flat sheet)

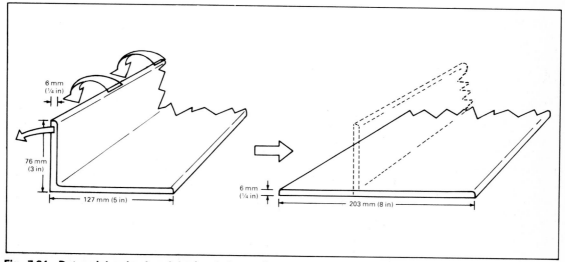

Fig. 7.24 Determining load weight (angled metal)

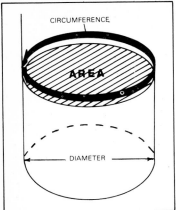

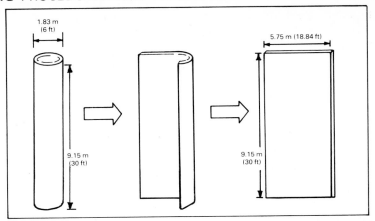

Fig. 7.25 Calculating the area and circumference of a circle

Fig. 7.26 Determining load weight (cylinder)

DETERMINING LOAD WEIGHTS

The most important step in any lifting operation is the determination of the weight of the load to be lifted. If this information cannot be obtained from shipping papers, design plans, catalogue data or from other dependable sources, it may be necessary to calculate the weight.

On erection plans, the size of steel beams or other sections is usually supplied together with their weight per unit length and the length of the section, so it is easy to compute the weight of any such loads which have to be lifted. Where angle, plate, or built-up members are involved, however, the weights must be calculated, and the memorisation of one basic weight and two formulae will give a reasonably accurate estimation of the weight.

The basic weight is that of 1 sq ft of steel an inch thick and is approximately 40 lb. Thus, two plates of steel each measuring 1½ in × 3 ft × 6 ft (Fig. 7.23) would weigh a total of

$$2 \times 1\tfrac{1}{2} \times 3 \times 6 \times 40 = 2160 \text{ lb}$$

In metric dimensions an equivalent steel plate 300 mm × 300 mm × 25 mm thick would weigh approximately 17 kg, and two plates 900 mm × 2000 mm × 40 mm thick would weigh 1130 kg.

The weights of steel angles also can be approximated with close enough results for safe job use. An angle is a structural section which can be considered as a bent plate with some additional metal at the heel for strength and a lesser amount of metal at the toes for ease in rolling. If the angle is flattened out, a plate results (Fig. 7.24). For example, a 5 in by 3 in by ¼ in angle would flatten out to approximately an 8 in by ¼ in plate. This should weigh 40 lb × 8/12 × ¼, or 6.65 lb per ft.

Weights of any structural shape can be computed in this manner by separating the parts or flattening them into rectangles which, in turn, become parts or multiples of a square foot of steel an inch thick.

Plates, however, are often rolled into cylindrical or other shapes which must be handled. It is necessary to determine the area of such parts before the weight can be calculated. This necessitates learning the two simple formulae for computing the circumference and the area of a circle. The circumference, or distance around the edge of a circle (Fig. 7.25), is found by multiplying the diameter by the number 3.14.

A tube 6 ft in diameter would have a circumference of 6 ft × 3.14, or 18.84 ft. To compute the weight of this tube, if it were 30 ft long and made of ⅜ in plate, mentally unroll it and flatten it out (Fig. 7.26). This gives a plate 18.84 ft wide by 30 ft long by ⅜ in thick, the weight of which is:

$$18.84 \times 30 \times \tfrac{3}{8} \times 40 = 8478 \text{ lb (3850 kg)}$$

The second formula gives the area of circular objects, as follows:

$$\text{Area} = 3.14 \times \frac{\text{diameter}}{2} \times \frac{\text{diameter}}{2}$$

Thus, if the tube had an end cap ⅜ in thick and 6 ft in diameter (Fig. 7.27) it would have a surface area:

Area = $3.14 \times \frac{6}{2} \times \frac{6}{2}$ = 28.3 ft² (2.63 m²)

and would weigh:

$28.3 \times \frac{3}{8} \times 40$ = 425 lb (193 kg)

For other material, using Imperial units, the weights are normally based on their weight per cubic foot, so you have to determine how many cubic feet of material (the volume) you are lifting in order to estimate the weight. In the metric system the weights are in kilograms per cubic metre (kg/m³).

For example, suppose you have a bundle of spruce timber to lift and the bundle is 12 ft long, 3 ft high and 4 ft wide (Fig. 7.28). The weight per cubic foot from Table 7.2 is 28 lb (or 448 kg/m³), so the weight of this bundle is $12 \times 3 \times 4 \times 28$ = 4032 lb (or 1830 kg).

The time taken to calculate the approximate weight of any object, whether steel plates, columns, girders, castings, bedplates, etc. is time well spent and may save a serious accident through failure of lifting gear. Tables 7.1 and 7.2 giving weights of various materials should enable the approximate weight of a given load to be calculated. When in doubt, do not hesitate to seek advice from an engineer or foreman on the job. Do not take a chance where the safety of men is involved.

CENTRE OF GRAVITY (C of G)

It is always important in a lifting operation to rig the load so that it is stable. A stable load is one in which its centre of gravity is directly below the main hook and below the lowest point of attachment of the slings. (Fig. 7.29)

The centre of gravity of an object is that point at which the object will balance. The entire weight may be considered as concentrated at this point. A suspended object will always move so that its centre of gravity is below the point of support. In order to make a level or stable lift, the crane or hook block must be directly above this point. Thus a load which is slung above and through its centre of gravity will be stable and will not tend to topple or slide out of the slings. (Fig. 7.30)

On an object of uniform shape or composition there is no problem in determining where the centre of gravity lies since it is at the centre of the shape or object, but on odd-shaped bodies where it cannot be determined easily, the lifting crew must judge where it lies, to try to lift with the hook over that point and then correct it by trial and error methods, moving the hook, load and sling suspension a little at a time until a satisfactory result is obtained. The object will usually tilt until its centre of gravity is directly beneath the load hook, so there is an indication of the direction in which to shift the slings. Remember too, that when a load's centre of gravity is closer to one point of sling attachment than to the other, the sling legs will have to be of unequal length which means that their angles and loads will also be unequal.

If a load tilts when it is lifted and this is not corrected then one leg of the sling will see a large load increase and the load on the other will decrease. If any load tilts more than 5° after it is lifted clear of the ground it should be landed and re-slung.

It is equally important to ensure that the points of support of a load (i.e., where the slings are attached to the load) lie above the centre of gravity and not below it, for the centre of gravity will always tend to move to the lowest point possible below the point of support. This precaution is especially applicable when lifting with pallets, skids or the base of any object since all these have a tendency to topple. The greatest stability will be achieved when the sling angles are very much larger than the angle formed between the plane of

TABLE 7.1

APPROXIMATE WEIGHTS ROUND STEEL BARS AND RODS kg per metre/lb per foot							
Diameter		Weight		Diameter		Weight	
mm	in	kg/m	lb/ft	mm	in	kg/m	lb/ft
5	3/16	.140	.094	35	1⅜	7.51	5.05
6	¼	.250	.167	38	1½	8.95	6.01
8	5/16	.390	.261	40	1⅝	10.50	7.05
9	⅜	.560	.376	44	1¾	12.18	8.18
11	7/16	.760	.511	48	1⅞	13.98	9.39
13	½	.99	.668	51	2	15.88	10.68
14	9/16	1.26	.845	54	2⅛	17.92	12.06
16	⅝	1.55	1.04	56	2¼	20.15	13.52
19	¾	2.24	1.50	60	2⅜	22.40	15.06
22	⅞	3.04	2.04	64	2½	24.80	16.69
26	1	3.97	2.67	67	2⅝	27.40	18.40
28	1⅛	5.04	3.38	70	2¾	30.00	20.20
30	1 3/16	5.61	3.77	73	2⅞	32.90	22.07
32	1¼	6.20	4.17	76	3	35.80	24.03

182 OPERATING PROCEDURES AND PRECAUTIONS

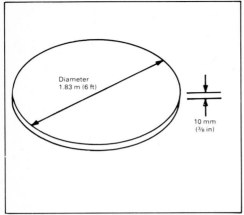

Fig. 7.27 Determining load weight (circular disc)

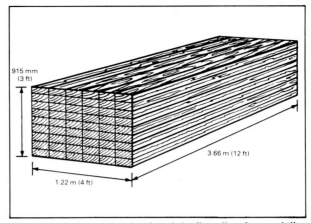

Fig. 7.28 Determining load weight (bundle of material)

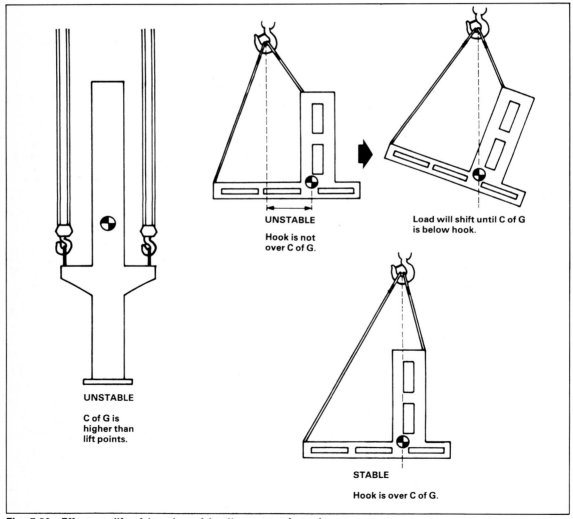

Fig. 7.29 Effect on lift of location of load's centre of gravity

TABLE 7.2

WEIGHTS OF MATERIALS (based on volume)

Material	Approximate weight kg/m^3	lb/ft^3	Material	Approximate weight kg/m^3	lb/ft^3
METALS			**TIMBER, AIR-DRY**		
Aluminium	2643	185	Cedar	350	22
Brass	8570	535	Fir, Douglas, seasoned	545	34
Bronze	8010	500	Fir, Douglas, unseasoned	640	40
Copper	8970	560	Fir, Douglas, wet	800	50
Iron	7690	480	Fir, Douglas, glue laminated	545	34
Lead	11374	710	Hemlock	480	30
Steel	7850	490	Pine	480	30
Tin	7370	460	Poplar	480	30
MASONRY			Spruce	448	28
Ashlar masonry	2240–2560	140–160	**LIQUIDS**		
Brick masonry, soft	1760	110	Alcohol, pure	785	49
Brick masonry, common (about 3 tonnes per thousand)	2000	125	Petroleum	670	42
Brick masonry, pressed	2240	140	Oils	930	58
Clay tile masonry, average	960	60	Water	993	62
Rubble masonry	2080–2480	130–155	**EARTH**		
Concrete, cinder, haydite	1600–1760	100–110	Earth, wet	1600	100
Concrete, slag	2080	130	Earth, dry	1200	75
Concrete, stone	2307	144	Sand and gravel, wet	1920	120
Concrete, stone, reinforced	2400	150	Sand and gravel, dry	1680	105
ICE AND SNOW			River sand	1920	120
Ice	900	56	**VARIOUS BUILDING MATERIALS**		
Snow, dry, fresh fallen	128	8	Cement, portland, loose	1505	94
Snow, dry, packed	190–400	12–25	Cement, portland, set	2930	183
Snow, wet	430–640	27–40	Lime, gypsum, loose	850–1025	53–64
MISCELLANEOUS			Mortar, cement-lime, set	1650	103
Asphalt	1280	80	Crushed rock	1440–1760	90–110
Tar	1200	75			
Glass	2560	160			
Paper	960	60			

support and the line through the centre of gravity, but this type of load will never be as stable as one in which the sling attachments are above the centre of gravity. (Fig. 7.31)

SIGNALLING

Slingers or banksmen may be frequently required to act as signalmen for crane operators and there are several precautions and procedures that should be observed in these operations.

Whenever the operator is obstructed in his view of the path of travel of any part of the equipment, its load or components, a competent signalman is required to be stationed:

- In full view of the operator.
- With a full view of the intended path of travel of the equipment, load or components, yet clear of the intended path of travel and should assist the operator by keeping that part of the equipment under observation when it is out of view of the operator and by communicating with the operator by the use of pre-arranged visual signals or a suitable communication system.

The signalman must:

- Be fully qualified by experience with the operation.
- Wear high-visibility gloves.
- Use hand signals only when conditions are such that his signals are clearly visible to the operator.
- Be made responsible for keeping the public and all unauthorised personnel outside the crane's operating radius.
- Direct the load so that it never passes over anyone.

OPERATING PROCEDURES AND PRECAUTIONS

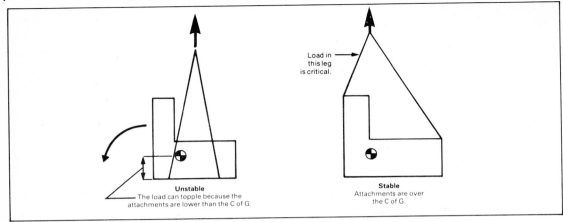

Fig. 7.30 Effect on lift of location of load's centre of gravity

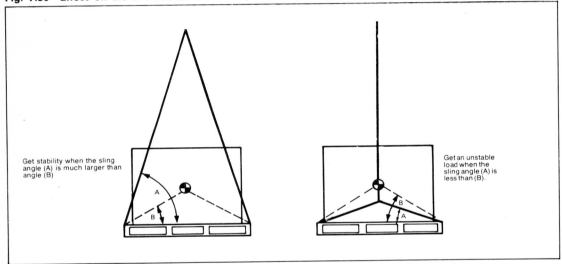

Fig. 7.31 Relationship of load stability with load's centre of gravity and sling angle

This signalman must be in constant communication with the crane operator, either visually with hand signals or audibly by radio throughout the operation. If the operator loses contact with the signalman for any reason, he must stop the movement of the crane until communication is restored. Where loads are picked up at one point and lowered at another, two signalmen may be required, one to direct the lift and one to direct the descent.

If it is desired to give instructions to the operator, other than provided for by the established signal system, the operator should be instructed to first stop all lifting motions.

Hand signals can be used effectively when the distance between the operator and the signalman is not great, but two-way radios should be used when the distance or atmospheric conditions prevent clear visibility. Adequate lighting and signalling arrangements must be available during night work and the equipment must not be operated when either is inadequate.

When two or more cranes or hoists are used to lift one load, one qualified person should be responsible for the operation. He should analyse the operation and instruct all personnel involved as to the correct positioning and slinging of the load, and the movement to be made. He should also ensure that the operators of the units and the signalman (or men) are able to communicate with each other via audio systems.

OPERATING PROCEDURES AND PRECAUTIONS

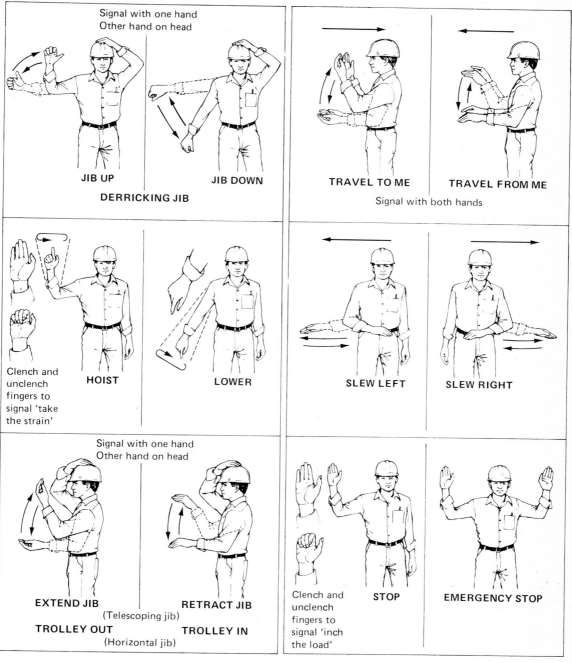

THE SIGNALLER (BANKSMAN) MUST BE IN A SAFE POSITION FROM WHERE HE CAN BE SEEN CLEARLY BY THE CRANE OPERATOR AND FROM WHICH HE CAN SEE THE CRANE LOAD THROUGHOUT THE LIFTING OPERATION, IF THIS IS PRACTICABLE. THE SIGNALLER SHOULD FACE THE CRANE OPERATOR.

ADDITIONAL COMPETENT SIGNALLERS MUST BE POSTED IF OBSTRUCTIONS PREVENT THE LOAD BEING SEEN CONTINUOUSLY BY A SINGLE SIGNALLER DURING PART OF THE LIFTING OPERATION.

Fig. 7.32 Hand signals for controlling crane operations

Appendix A

BRITISH STANDARDS

CRANES

Derrick, Power-driven	BS 327
Derrick, Mobile/Tower, Safe Use	CP 3010
Safe Use of Cranes (complements CP 3010)	BS 5744
Mobile, Power-driven	BS 1757
Tower, Power-driven	BS 2799

LIFTING TACKLE

Wire Ropes for Cranes, etc.	BS 302
Wire Rope Slings	BS 1290
Round Steel Wire for Wire Ropes	BS 2763
Eye Terminations for Wire Ropes	BS 5281
Ropes, Manila, Natural Fibre	BS 2052
Ropes, Man-made Fibre	BS 4928
Chain Slings, Alloy Steel	BS 3458
Flat Lifting Slings, Wire Coil	BS 3481 Part 1
Flat Lifting Slings, Man-made Fibre	BS 3481 Part 2
Eyebolts for Lifting	BS 4278
Shackles for Lifting	BS 3551
Swivels for Lifting	BS 4283
Pulley Blocks	BS 4536
Bolts, Unified	BS 1768/9
Bolts, Metric	BS 3692:4190

METRICATION

Conversion Factors	BS 350
The Metric System in the Construction Industry	PD 6031

PRINCIPAL STATUTORY REQUIREMENTS FOR CONSTRUCTION SITE CRANES IN THE UNITED KINGDOM

1. Health & Safety at Work etc. Act 1974
   Principally sections 2, 3, 4, 6 and 7
2. Construction (Lifting Operations) Regulations 1961
3. Construction (General Provisions) Regulations 1961
4. Construction (Working Places) Regulations 1966
5. Electricity Regulations 1908 and 1944

Appendix B

ADDRESSES

HEALTH AND SAFETY INSPECTORATE AREA OFFICES

SOUTH WEST
Avon, Cornwall, Devonshire,
Gloucestershire, Somersetshire,
Isles of Scilly:
Inter City House, Mitchell Lane,
Victoria Street, Bristol BS1 6AN.

SOUTH
Berkshire, Dorsetshire,
Hampshire, Isle of Wight:
Priestley House, Priestley Road,
Basingstoke, Hants. RG24 9NW.

SOUTH EAST
Kent, Surrey, East/West Sussex:
Paymaster General's Buildings,
Russell Way, Crawley,
W. Sussex RH10 1UH

LONDON – NORTH WEST
Chancel House,
Neasden Lane, London NW10 2UD.

LONDON – NORTH EAST
Royal London House,
18 Finsbury Square,
London EC2 1DH.

LONDON – SOUTH
1 Long Lane,
Southwark, London SE1 4PG.

EAST ANGLIA
Essex, Norfolk, Suffolk:
39 Baddow Road,
Chelmsford, Essex CM2 0HL.

NORTHERN HOME COUNTIES
Bedfordshire, Buckinghamshire,
Cambridgeshire, Hertfordshire:
6th Floor, King House,
George Street West,
Luton, Beds. LU1 2DD.

EAST MIDLANDS
Leicestershire, Oxfordshire,
Northamptonshire, Warwickshire:
5th Floor, Belgrave House,
1 Greyfriars,
Northampton NN1 2LQ.

WEST MIDLANDS
West Midlands:
McLaren Buildings,
2 Masshouse Circus, Queensway,
Birmingham B4 7NP.

WALES
Clwyd, Dyfed, Gwent, Gwynedd,
Mid/South/West Glamorgan, Powys:
Brunel House, 2 Fitzalan Road,
Cardiff CF2 1SH.

MARCHES
Staffordshire, Shropshire,
Herefordshire, Worcestershire:
Norwich Union House,
40 Trinity Street, Hawley,
Stoke-on-Trent, Staffs.

NORTH MIDLANDS
Derbyshire, Lincolnshire,
Nottinghamshire:
Birkbeck House,
Trinity Square,
Nottingham NG1 4AU.

SOUTH YORKSHIRE, HUMBERSIDE
South Yorkshire, Humberside:
Sovereign House,
40 Silver Street,
Sheffield, S. Yorks. S1 2ES.

WEST/NORTH YORKSHIRE
West/North Yorkshire:
8 St. Paul's Street,
Leeds, W. Yorks. LS1 2LE.

GREATER MANCHESTER AREA
Greater Manchester:
Quay House, Quay Street,
Manchester M3 3JB.

MERSEYSIDE
Merseyside, Cheshire:
The Triad, Stanley Road,
Bootle, Merseyside L20 3PG.

NORTH WEST
Lancashire, Cumbria:
Victoria House,
Ormskirk Road,
Preston, Lancs. PR1 1HH.

NORTH EAST
Tyne/Wear, Northumberland,
County Durham, Cleveland:
Government Buildings,
Kenton Bar,
Newcastle-upon-Tyne NE1 2YX.

SCOTLAND EAST
Borders, Lothian, Central,
Tayside, Grampian, Highland,
Fife, Orkney/Shetland Isles:
Meadowbank House,
153 London Road,
Edinburgh EH8 7AU.

SCOTLAND WEST
Dumfries/Galloway,
Strathclyde, Western Isles:
314 St. Vincent Street,
Glasgow G3 8XG.

EMPLOYMENT MEDICAL ADVISORY SERVICE REGIONAL OFFICES

Enquire at local HSI offices.

HM STATIONERY OFFICE GOVERNMENT BOOKSHOPS

LONDON
Callers only:
49 High Holborn,
London WC1V 6HB.

Trade/London area
mail orders:
PO Box 569,
London SE1 9NH.

EDINBURGH
13a Castle Street,
Edinburgh EH2 3AR.

CARDIFF
41 The Hayes,
Cardiff CF1 1JW.

MANCHESTER
Brazennose Street,
Manchester M60 8AS.

BRISTOL
Southey House,
Wine Street,
Bristol BS1 2BQ.

BIRMINGHAM
258 Broad Street,
Birmingham B1 2HE.

BELFAST
80 Chichester Street,
Belfast BT1 4JY.

BRITISH STANDARDS INSTITUTION

2 Park Street,
London W1A 2BS.

Sales:
101 Pentonville Road,
London N1 9ND.